Student Study Guide

to accompany

Microbiology
A Human Perspective

Second Edition

Eugene W. Nester, C. Evans Roberts
University of Washington

Nancy N. Pearsall

Denise G. Anderson
University of Washington

Martha T. Nester

Prepared by
William D. O'Dell
University of Nebraska-Omaha

Boston Burr Ridge, IL Dubuque, IA Madison, WI New York San Francisco St. Louis
Bangkok Bogotá Caracas Lisbon London Madrid
Mexico City Milan New Delhi Seoul Singapore Sydney Taipei Toronto

WCB/McGraw-Hill

A Division of The McGraw·Hill Companies

Student Study Guide to accompany
MICROBIOLOGY: A HUMAN PERSPECTIVE, SECOND EDITION

This book is printed on acid-free paper.

6 7 8 9 0 QPD/QPD 9 3 2 1 0

ISBN 0-697-28605-3

www.mhhe.com

Contents

Preface

You are about to embark on a new adventure, the study of microbiology. Some of you may be anxious about starting because you are not sure what lies ahead. Some may be anxious because they do know what is ahead. What lies ahead are new ideas and a fascinating world of microorganisms. You will discover or rediscover the impact that things too small to be seen without a microscope can have on your daily lives. Hopefully, you will marvel at the fact that something that may not even be alive, and is only a millionth of an inch big, can cause a grown man or woman to cough and sneeze for a week! Hopefully, you will grow excited at the complexity and beauty of your own immune system. Maybe you won't care and you will just want to finish the course. Either way you are starting a new adventure and this can be your guide.

This study guide is not intended to replace your textbook, but instead to supplement the textbook and to help direct you through the concepts and terminology of microbiology. The Student Study Guide follows the organization of your textbook. Following a short overview of the chapter contents, each chapter begins with a section titled, **KEY CONCEPTS**. This section states in numbered sentences the major concepts from the chapter in a direct, applied and usually nontechnical fashion. As you begin to study a chapter, and usually before the material is presented in class, you should read these concepts because they introduce you to the chapter.

The next section is the **SUMMARY**. This is an outline of the topics in the chapter and should help you in your reading and understanding of how ideas and concepts fit together and are related. Before you begin to read the chapter, it would be a good idea to browse through the summary to help organize your own thoughts. The summary is also helpful in taking reading notes as you proceed with your reading assignments. Use it to fill in details from your reading and from your class. It is usually very helpful if you have read the key concepts and summary before going to a lecture class. In addition, you will probably find it helpful to page through the textbook and identify the terms in bold lettering before you go to lecture. When you hear the terms in class they will not sound so foreign if you have already seen them once or twice.

The new terminology associated with microbiology is often one of the most difficult aspects of learning this or any new subject. To help you with this, the section **VOCABULARY: TERMS AND DEFINITIONS** provides you with definitions that are directly or closely paraphrased from your textbook. These terms appear in bold lettering in the text. Practice with these terms should help you master their meanings. When you have finished your reading, you can test your knowledge in the **SELF-TEST OF READING MATERIAL**. This section offers you the opportunity through multiple choice or matching questions to see if you have comprehended the reading material. The questions typically reflect the major concepts and sections of the chapter. Answers are provided so you may check immediately to see how you did.

The final section, **REVIEW QUESTIONS**, includes written questions that ask you simply to recall some text material or to think about something and synthesize a new answer. You can derive the most benefit from this section if you write out the answers because you will use the terms and arrange ideas into a specific answer. Hopefully, you will even find some of the questions fun and interesting.

Regardless of your reasons or approach to studying microbiology, enjoy yourself.

1 MICROBIOLOGY IN THE BIOLOGICAL WORLD

This chapter provides an introduction to microbiology with a discussion of significant milestones that have been important in the development of microbiology. It also defines new directions for microbiologists of the future. The concepts of cell theory and cell types are introduced. The chapter concludes with a description of the various members of the microbial world and how we name them.

KEY CONCEPTS

1. Microorganisms have determined the course of history because of the diseases they cause.
2. New infectious diseases appear as lifestyles change, people travel to exotic places, and techniques for growing and identifying organisms and viruses improve.
3. Cells are the basic units of life, and all cells must carry out the same critical functions in order to survive.
4. Two major types of cells exist: the prokaryotes, which do not contain a "true" nucleus or other membrane-bound internal structures, and the eukaryotes, which do contain a true nucleus.
5. The microbial world consists of prokaryotes and eukaryotes, as well as nonliving agents, the viruses, viroids and prions.
6. All prokaryotes can be divided into two very distinct groups, the Eubacteria and the Archaea, based on their chemical composition.

SUMMARY

I. Introduction
   A. Microorganisms are, in large part, responsible for determining the course of human history.
   B. The use of modern sanitation facilities, vaccinations, as well as antibiotics has dramatically reduced the incidence of infectious disease.

II. Microorganisms Discovered
   A. Anton van Leeuwenhoek discovered microorganisms over 300 years ago by viewing water samples through lenses that magnified three hundred fold.
   B. The theory of spontaneous generation was revived with the discovery of the microbial world.
      1. Differing results from different investigators led to the controversy of whether living organisms could arise from dead organic matter. This controversy was not resolved until the 1860s.
      2. Pasteur demonstrated that the air is filled with microorganisms and showed that swan-necked flasks containing sterile infusions could remain sterile indefinitely.
      3. Tyndall and, independently, Cohn discovered that heat-resistant forms of bacteria, or endospores, were present in certain infusions.

III. Medical Microbiology - Past Triumphs
   A. Between 1875 and 1918, most disease-causing bacteria were identified.

IV. Medical Microbiology - Future Challenges
   A. "New" diseases are appearing. These include Legionnaires' disease, toxic shock syndrome, Lyme disease, AIDS, and hantavirus disease.
   B. Many diseases that were on the wane are now increasing in frequency. These include mumps, whooping cough, diphtheria, and most recently, tuberculosis.
   C. Organisms are becoming increasingly resistant to antibiotics.

V. Beneficial Applications of Microbiology - Past and Present
   A. Human life could not exist without the activity of microorganisms.
   B. Microorganisms have been used for centuries for food production.

VI. Biotechnology - New Applications for Microorganisms
    A. Microorganisms are now being developed to produce vaccines, clean up the environment, and to carry out many other processes designed to make life more comfortable.

VII. Cell Theory
    A. Schleiden and Schwann in the mid-1800s proposed the cell theory - that cells are the basic units of life.

VIII. Similarity in Composition and Function of All Cells
    A. All cells growing independently of other cells, such as bacteria, have one basic function - to reproduce. To do this, they must generate energy and synthesize the components of living matter.
    B. All cells are composed of the same macromolecules, such as nucleic acids (DNA and RNA) and proteins, which, in turn, are composed of the same subunits.

IX. Basic Cell Types
    A. There are two cell types: prokaryotic and eukaryotic. Prokaryotic cells are simple, without membrane-bound internal structures. Eukaryotic cells are larger, more complex, and have several internal membrane-bound structures. All bacteria are prokaryotic; algae, fungi, and protozoa are eukaryotic.
    B. A cell type has been found that appears to be intermediate between prokaryotes and eukaryotes. It has a true nucleus but no mitochondria.
    C. Bacteria can be divided into two domains, the Eubacteria and the Archaea.
        1. Both groups are similar microscopically, but differ in the chemical composition of several structures.
        2. The two groups are not closely related to each other or to the eukaryotes.
    D. Eubacteria include bacteria most familiar to microbiologists.
        1. This group is very diverse.
        2. The "typical" bacteria, which include both Eubacteria and Archaea, are the most common and although they are also heterogeneous, they do share some obvious properties.
    E. Archaea often grow under extreme conditions of temperature and salinity.

X. Members of the Microbial World
    A. The members include all unicellular organisms, which includes all prokaryotes.
    B. Algae, fungi, and protozoa are the eukaryotic members of the microbial world.
    C. Viruses, viroids, and prions are nonliving members of the microbial world.

XI. Nomenclature of Organisms
    A. All organisms are named according to the binomial system of genus and species.
    B. Names and descriptions of most bacteria are published in *Bergey's Manual of Systematic Bacteriology*.

VOCABULARY: TERMS AND DEFINITIONS

The following list contains new terms introduced in this chapter. Use these terms to fill-in the blanks of the sentences that follow and you will have a definition or description of each term.

microorganisms	cell theory	fungi
viruses	prokaryotic	protozoa
darkfield illumination	algae	prions
endospores	eukaryotic	
bioremediation	binomial	

1. _____ is a technique of microscopy in which faint objects appear brightly lit against a dark background.

2. Eukaryotic members of the microbial world include the _____ , the _____ , and the _____ .

3. Organisms that can be seen only with the aid of a microscope are called _____ .

4. _____ is a process by which living microorganisms are used to help clean up the environment.

5. _____ are heat-resistant, nondividing forms of bacteria.

6. _____ appear to be proteins without nucleic acid, either DNA or RNA.

7. _____ consist of a piece of genetic material surrounded by a protective protein coat.

8. The system developed by Linnaeus for naming organisms is known as the _____ system of nomenclature.

9. The _____ states that all organisms are composed of cells and that cells are the fundamental units of life.

10. Cells that do not have a membrane surrounding their genetic material are known as _____ cells.

11. Cells that have a membrane surrounding their genetic material are known as _____ cells.

SELF-TEST OF READING MATERIAL

1. Leeuwenhoek's discoveries were significant because
   a. he was the first person to use a microscope.
   b. he carefully recorded and reported his results.
   c. of his finely ground lenses.
   d. he was a renowned scientist of his day.
   e. he limited his observations.

2. Infectious diseases have essentially been eliminated by antibiotics and vaccination and are no longer a public health problem.
   a. True
   b. False

3. Diseases, once thought to be controlled, which are reappearing in developed countries include
   1. measles
   2. whooping cough
   3. AIDS
   4. tuberculosis
   5. mumps

   a. 1,2,3,4
   b. 2,3,4,5
   c. 3
   d. 1,2,4,5
   e. 1,3,4,5

4. The number of disease-producing microorganisms is _____ of the total number of microorganisms.
   a. a very large proportion
   b. about half
   c. a very small proportion

5. The discovery of the microbial world by Leeuwenhoek created a controversy that lasted almost 300 years. This controversy centered on
   a. the causes of disease.
   b. spontaneous generation.
   c. flies and maggots.
   d. the church and the scientist.
   e. the origin of life.

6. Pasteur's experiments demonstrated that microorganisms in the _____ were indistinguishable from those that grew in contaminated flasks.
   a. corks
   b. broth
   c. air
   d. water
   e. soil

7. Tyndall and, independently, Cohn discovered a heat-resistant form of bacteria. This bacterial form became known as a(n)
   a. exospore.
   b. vegetative cell.
   c. prokaryotic cell.
   d. cell wall.
   e. endospore.

8. The cell theory states that
   a. cells come from other cells.
   b. cells are the fundamental units of life.
   c. all organisms are composed of cells.
   d. Only b. and c. are correct
   e. a., b., and c. are correct.

9. The original distinction between the two types of cells, prokaryotic and eukaryotic, was made on the basis of the
   a. structure of the cell wall.
   b. absence or presence of mitochondria.
   c. absence or presence of a nuclear membrane.
   d. absence or presence of ribosomes.
   e. structure of the cell membrane.

10. Bacteria, with the exception of Archaea, are prokaryotic cells.
    a. True
    b. False

11. While the Eubacteria, the "typical" bacteria, are a heterogeneous group, they do share some features in common. Which of the following features are shared by the Eubacteria?
    1. single-cell prokaryotes
    2. rigid cell walls
    3. single-cell eukaryotes
    4. multiply by binary fission; one cell divides into two
    5. rigid cell walls with cellulose
    a. 1,2,4
    b. 2,3,4
    c. 3,4,5
    d. 1,2
    e. 3,5

12. Which of the following would probably not be included in the microbial kingdom?
    a. all single-cell organisms
    b. multicellular organisms with a cellular level of organization
    c. organisms with tissues and organs
    d. all single-cell eukaryotes
    e. all prokaryotes

13. We cannot describe viruses as prokaryotes or eukaryotes because viruses
    a. are not living.
    b. are agents and not organisms.
    c. were discovered after prokaryotes and eukaryotes.
    d. are not cellular.
    e. do not contain any genetic information.

14. Viruses are too small to contain all of the machinery and molecules necessary for life.
    a. True
    b. False

15. Which of the following is a correct way to represent the scientific name for a specific bacterium?
    a. <u>Bacillus</u> <u>cereus</u>
    b. <u>Bacillus</u> <u>Cereus</u>
    c. bacillus cereus
    d. Bacillus cereus
    e. bacillus Cereus

REVIEW QUESTIONS

1. What is the argument for including multicellular organisms such as algae and mushrooms in the microbial kingdom?

2. What problems contributed to the confusion of the early investigations into spontaneous generation?

3. Describe a few ways in which biotechnology might improve our lives now and in the future.

ANSWERS:
Vocabulary - Terms and Definitions
1. darkfield illumination 2. algae, fungi, protozoa 3. microorganisms 4. bioremediation 5. endospores
6. prions 7. viruses 8. binomial 9. cell theory 10. prokaryotic 11. eukaryotic
Self-Test of Reading Material
1. b 2. b 3. d 4. c 5. b 6. c 7. e 8. e 9. c 10. b 11. a 12. c 13. d 14. a 15. a

NOTES

2 BIOCHEMISTRY OF THE MOLECULES OF LIFE

To understand how microorganisms live and die, produce disease, and do all of the other amazing things they do, requires some working knowledge of chemistry. This chapter presents the fundamental concepts of the chemistry of living organisms. It starts with the atom, the simplest level of organization, and moves to higher levels, to finish with the macromolecules. The structure and function of proteins, polysaccharides, nucleic acids and lipids are described.

KEY CONCEPTS

1. Four elements, carbon, oxygen, hydrogen, and nitrogen make up over 98% of all living matter. Two other elements, phosphorus and sulfur, are also very important.
2. The bonds that hold atoms together result from electrons interacting with each other. Bonds vary in strength which gives molecules characteristic properties.
3. Weak bonds are important in biological systems, since they often determine the most important properties of the molecules and are responsible for their proper functioning.
4. All life is based on the bonding properties of water which comprises over 90% of the cell's weight.
5. Macromolecules consist of many repeating subunits, each subunit consisting of a small, simple molecule. The subunits are synthesized, then bonded to form the macromolecule.

SUMMARY

I. Elements and Atoms
   A. An element is a pure substance that consists of a single type of atom. Atoms are the basic units of all matter. They consist of three major components: electrons, protons, and neutrons.

II. Formation of Molecules: Chemical Bonds
   A. Chemical bonds are of two types: strong and weak. The stronger the bond, the more energy is required to break it.
   B. Strong bonds are usually covalent bonds formed when atoms share electrons to fill their outer shell and thereby achieve maximum stability.
      1. Covalent bonds vary in their distribution of shared electrons, which results in the molecule having a positive and negative charge at different sites.
   C. Ionic bonds are formed by the loss and gain of electrons between atoms. In aqueous solutions, they are weak.
   D. Hydrogen bonds are weak but biologically very important. They hold the two strands of DNA together and are important in determining the shape of proteins.
      1. They result from the attraction of positively charged H atoms to negatively charged N or O atoms.

III. Important Molecules of Life
   A. Small molecules in the cell are both organic and inorganic.
      1. The inorganic molecules include many that are required for enzyme function.
      2. Organic molecules are mainly compounds that are being metabolized or molecules that are the subunits of macromolecules.
   B. All very large molecules in the cell (macromolecules) consist of repeating subunits called monomers.
      1. There are three important macromolecules.
         (1). Proteins are chains of amino acids that form a polypeptide.
         (2). Polysaccharides are chains of monosaccharides that form branching structures.
         (3). The nucleic acids are DNA and RNA, which are chains of nucleotides.
   C. Proteins are polymers of amino acids.
      1. Amino acids consist of a molecule with a carboxyl group and an amino group bonded to the same carbon atom. The carbon atom is bonded, in turn, to a side chain. Twenty different amino acids, each differing in their side chains, are present in proteins.

2. Peptide bond synthesis: A covalent bond is formed between the amino group of one amino acid and the carboxyl group of the adjacent amino acid with the removal of HOH (dehydration synthesis).
3. Levels of protein structure: Three features characterize a protein.
   (1). Its primary structure
   (2). Its shape, whether globular or long fibers
   (3). Whether or not the protein consists of one or several polypeptide chains. A variety of weak bonds are involved in maintaining the three dimensional shape of proteins.
4. Substitute proteins contain covalently bonded molecules other than amino acids. These include glycoproteins (sugars) and lipoproteins (lipids).

D. Polysaccharides are polymers of monosaccharide (carbohydrate) subunits. Carbohydrates (sugars) contain a large number of alcohol groups (-OH) in which the C atom is also bonded to an H atom to form H-C-OH.
   1. Monosaccharides are the subunits of polysaccharides. The most common are hexoses (6C) and pentoses (5C).
   2. Disaccharides are two monosaccharides joined together with a loss of water (dehydration synthesis).
   3. Different polysaccharides vary in size, their degree of branching, the bonding of monosaccharides to one another and the monosaccharides involved.

E. Nucleic acids, which include deoxyribonucleic acid (DNA) and ribonucleic acid (RNA), are polymers of nucleotide subunits and are unbranched.
   1. DNA. The nucleotides are composed of three units: a nitrogen base, [purine (adenine or guanine) or pyrimidine (thymine or cytosine)], covalently bonded to deoxyribose, which, in turn, is bonded to a phosphate molecule. A phosphate bonded to the sugar molecule joins the nucleotides together. DNA occurs in the cell as a double-stranded helix in which the two strands are held together by hydrogen bonding between adenine and thymine and between guanine and cytosine.
   2. Ribonucleic acid (RNA). Nucleotides are the same as in DNA except that ribose replaces deoxyribose, and uracil replaces thymine. RNA is shorter in length and does not occur as a double helix. Three different types of RNA exist in the cell.

F. Lipids are biologically important but are too small and heterogeneous to be considered macromolecules.
   1. All lipids have one property in common, they are insoluble in water, but soluble in organic solvents. This difference in solubility is due to their nonpolar, hydrophobic nature.
   2. They are not composed of similar subunits but rather a variety of substances that differ in chemical structure.
   3. Simple lipids, which contain only C, H, and O, include fats and steroids. Fats are composed of glycerol covalently bonded to fatty acids. Steroids have a four-membered ring structure.
   4. Compound lipids contain fatty acids and glycerol and, in addition, often elements other than C, H, and O. They include phospholipids, lipoproteins, and lipopolysaccharides. They all play important roles in the cell envelope of bacteria.
   5. Phospholipids consist of two parts, each with different properties. One end is polar and therefore is soluble in water. The other end, containing only C and H, is nonpolar and therefore insoluble in water, but soluble in organic solvents.

VOCABULARY: TERMS AND DEFINITIONS

The following list contains new terms introduced in this chapter. Use these terms to fill-in the blanks of the sentences that follow and you will have a definition or description of each term.

neutrons	**macromolecules**	**fats**
electrons	**polymers**	**oils**
covalent bonds	**dehydration synthesis**	**phospholipids**
ionic bonds	**disaccharide**	
hydrogen bonds	**nucleotides**	

1. _____ result from the attraction of positively charged H atoms to negatively charged N or O atoms.

2. A strong bond made by sharing electrons between atoms is a _____ .

3. _____ are lipids that are liquids at room temperature.

4. Polymers are made from their monomer units by the process of _____ .

5. _____ are negatively charged particles that participate in the bonding of atoms.

6. Membrane molecules that have both hydrophobic and hydrophilic parts are known as _____ .

7. Sucrose is an example of a _____ .

8. Atomic particles found in the nucleus of an atom that do not have an electrical charge

   are called _____ .

9. _____ are the monomers or building blocks of nucleic acids.

10. Proteins, polysaccharides, and nucleic acids are examples of very large molecules called _____ .

11. The combination of fatty acids and glycerol which are solids at room temperature

    are known as _____ .

12. _____ involve the complete transfer of electrons from one atom to another.

13. Large molecules formed by joining together the same small molecules or monomers,

    are called _____ .

14. Protein molecules are held in their three-dimensional shape by _____ .

SELF-TEST OF READING MATERIAL

1. The sodium ion, Na^+, has a single positive charge because it has
   a. more neutrons in its nucleus than electrons in orbit.
   b. more electrons in its nucleus than protons in orbit.
   c. six electrons.
   d. more protons in its nucleus than electrons in orbit.
   e. more electrons in its nucleus than neutrons in orbit.

2. Atoms are electrically neutral. Which of the following has an electrical charge?
   1. protons
   2. electrons
   3. ions
   4. neutrons
   5. atoms

   a. 1,3,5
   b. 3,4,5
   c. 1,2,3
   d. 3
   e. 1,2,3,4,5

3. The number of protons in the nucleus of an atom is its
   a. atomic number.
   b. number of electron orbitals.
   c. atomic weight.
   d. molecular weight.
   e. valence.

4. Weak bonds that are responsible for holding the strands of DNA together are
   a. ionic bonds
   b. hydrogen bonds
   c. disulfide bonds
   d. nitrogen bonds
   e. water bonds

5. Biological molecules made by covalently bonding amino acids are called
    a. proteins.
    b. lipids.
    c. nucleic acids.
    d. disaccharides.
    e. polysaccharides.

6. All macromolecules share the common feature of being synthesized by joining subunits or monomers together. This joining process is known as
    a. hydrolysis.
    b. dehydration synthesis.
    c. hydrogen bonding
    d. hydration synthesis.
    e. dehydrolysis.

7. The number and sequence of amino acids determines the
    a. primary structure of a polysaccharide.
    b. secondary structure of a protein.
    c. the secondary structure of a polysaccharide.
    d. the primary structure of a protein.
    e. tertiary structure of a protein.

8. Amino acids contain which of the following functional groups?
    1. an NH_2 group
    2. an "R" group
    3. glycerol
    4. a COOH group
    5. fatty acid
        a. 1,2,4
        b. 3,5,
        c. 1,2,3
        d. 1,5
        e. 3,4,

9. Peptide bonds are found in ———————— while ester linkages are the bonds in ————————————— .
    a. lipids/proteins
    b. polysaccharides/nucleic acids
    c. proteins/lipids
    d. nucleic acids/polysaccharides
    e. proteins/polysaccharides

10. Lactose and sucrose are examples of
    a. DNA
    b. monosaccharides
    c. proteins
    d. polysaccharides
    e. disaccharides

11. The carbon to hydrogen to oxygen ratio of carbohydrates is
    a. very large.
    b. 1:2:1.
    c. 1:2:2.
    d. 2:1:4.
    e. impossible to determine

12. DNA differs from RNA in that DNA has
    1. two strands
    2. deoxyribose
    3. one strand
    4. ribose
    5. thymine
    6. uracil

        a. 1,2,4
        b. 3,4,6
        c. 2 only
        d. 1,2,5
        e. 1,2,6

13. Lipids that contain unsaturated fatty acids are usually solids at room temperature.
    a. True
    b. False

14. The backbone of the RNA molecule is composed of alternating units of
    a. ribose and phosphate.
    b. purine and pyrimidine.
    c. deoxyribose and phosphate.
    d. deoxyribose and ribose.
    e. uracil and ribose.

15. In the structure, C=O, how many pairs of electrons does carbon and oxygen share?
    a. one
    b. two
    c. three
    d. four
    e. five

16. Phospholipids
    a. have a polar and a nonpolar end.
    b. are found in cell membranes.
    c. function as cellular enzymes.
    d. are found only in prokaryotic cells.
    e. Two of the above are correct.

17. What type of bond is formed between the oxygen and the hydrogen atoms in a water molecule?
    a. nonpolar covalent bond
    b. ionic bond
    c. polar covalent bond
    d. savings bond
    e. hydrogen bond

Use the choices on the right to identify the molecules described on the left.

_____ 18. They are made of multiple (CH_2O) units.

_____ 19. They contain COOH and NH_2 groups.

_____ 20. They contain C, H, and O.

_____ 21. They have alternating sugar and phosphate groups.

_____ 22. They contain glycerol.

a. nucleic acids
b. proteins
c. lipids
d. carbohydrates
e. All of the above
f. None of the above

REVIEW QUESTIONS

1. What properties do macromolecules share?

2. Water is essential for life. What properties of water make it so critical for life? How are these properties related to the structure of water?

3. List the macromolecules in order of their importance to cells? Why is this so difficult?

NOTES

3 FUNCTIONAL ANATOMY OF PROKARYOTES AND EUKARYOTES

This chapter compares and contrasts the functional anatomy of prokaryotic and eukaryotic cell types. The emphasis is on the prokaryotic cell as represented by the bacteria. The functional anatomy consists of a description of the structure and function of various cellular components. Differences between prokaryotic and eukaryotic cellular structures serve as the basis for the selective action of certain antibiotics. Prokaryotic organisms also have several unique structures as well as a metabolically inactive endospore.

KEY CONCEPTS

1. The effectiveness of a microscope is based on its ability to visually separate two objects that are close together.
2. The many different microscopes that have been developed differ primarily in their lenses and their method of illuminating the specimen being studied.
3. All bacteria must be able to carry out the functions required for life and their small cells must therefore contain the structures to carry out these functions.
4. The composition of the rigid bacterial cell wall determines various properties of the organism, including its susceptibility to penicillin and its staining characteristics.
5. The cytoplasmic membrane largely determines what material gets into and out of the cell.
6. In bacteria, DNA occurs as a covalently closed, circular molecule without a surrounding membrane.
7. The structure of ribosomes differs in eukaryotic and prokaryotic cells; for this reason, some antibacterial agents such as streptomycin, kill bacteria but are harmless to human cells.
8. Some bacteria can develop into endospores, a type of cell that can survive adverse conditions such as high temperatures.

SUMMARY

I. Microscope Techniques: Instruments
   A. The simple microscope has one magnifying lens.
   B. The compound microscope has two sets of magnifying lenses, the objective lens and the ocular lens. The magnification is the product of the magnification of the individual lenses.
      1. Resolving power determines the usefulness of a microscope, and the amount of detail that can be seen. It depends on the wavelength of the illuminating material.
   C. The contrast between cells and surrounding medium must be increased to clearly see the bacteria.
   D. Types of microscopes
      1. Phase contrast microscope - a light microscope that increases the contrast between the bacteria and the surrounding medium.
      2. Interference microscope - the specimen is viewed in three dimensions.
      3. Darkfield microscope - the object being viewed is brightly illuminated against a dark background.
      4. Fluorescence microscope - objects that fluoresce are visualized with this light microscope.
      5. The electron microscope can resolve objects a thousand times better than the light microscope because it uses electrons rather than light to illuminate the specimen.
         a. Specimens must be prepared for viewing in the electron microscope. Such preparation includes, fixation, freeze-etching, and thin-sectioning.
         b. The scanning electron microscope sees surface details in a three-dimensional image.
         c. The scanning tunneling microscope also views surface structures of specimens but its resolving power is much greater than the electron microscope. Individual atoms can be seen.

II. Microscope Techniques: Staining
   A. Simple staining procedures use a single dye.
   B. Positive stains color cell components whereas, negative stains do not penetrate the cell.

    C. Differential staining techniques make use of a combination of dyes.
       1. The Gram stain divides bacteria into two groups based on their cell wall composition: gram-positive and gram-negative
       2. The acid-fast stain is used to identify members of the genus *Mycobacterium,* which cause tuberculosis and leprosy.
    D. Specific cell structures can be stained by using special techniques that depend on the structure of the material being stained.

III. Shapes of Bacteria
    A. Bacteria have three major shapes: cylindrical, or bacillus; spherical, or coccus; and spiral, or spirillum.
    B. Individual cells, especially cocci, may remain attached to one another forming particular arrangements that are useful in identifying genera.

IV. Functions Required in Prokaryotic Cells
    A. All cells must carry out the following functions:
       1. enclose the internal contents of the cell
       2. replicate genetic information
       3. synthesize cellular components
       4. generate, store, and utilize energy-rich compounds
    B. In addition, some cells can move, transfer genetic information, store reserve materials, and form a new cell type, the endospore.

V. Enclosure of Cytoplasm In Bacteria
    A. The cytoplasm is enclosed by three layers; the capsule, cell wall, and cytoplasmic membrane.
    B. Capsule or slime layer
       1. Some cells synthesize a capsule, or slime layer, when growing under certain nutritional conditions. Most capsules are composed of polysaccharides and are called the glycocalyx. Capsules are protective and also help attach bacteria to a variety of surfaces.
    C. Cell wall
       1. The bacterial cell wall has several important functions and its composition confers important properties to the cell. The cell wall holds a cell together and gives the cell its shape.
       2. The unique structure in all bacterial cell walls that confers rigidity is glycan, which consists of repeating subunits of N-acetylglucosamine joined to N-acetylmuramic acid. These subunits are joined to glycan subunits by amino acid bridges. Each layer is termed a peptidoglycan layer.
       3. Cell walls of gram-positive bacteria consist of multiple layers of peptidoglycan along with teichoic acid. Cell walls of gram-negative bacteria consist of one or a few layers of peptidoglycan with a complex outer membrane. This outer membrane is a phospholipid bilayer in which proteins called porins are embedded. Lipopolysaccharides project from the surface of the outer membrane. The peptidoglycan layer is located in the periplasm, the area between the outer membrane and the cytoplasmic membrane.
       4. The Archaea have no peptidoglycan in their cell walls and other bacteria have no cell wall.
       5. Differences in cell wall composition account for the gram-staining properties of the bacteria and their reaction to penicillin.
    D. Cytoplasmic membrane - All cells, both prokaryotic and eukaryotic, have a cytoplasmic membrane that determines which molecules enter and leave the cell.
       1. Chemical composition - The cytoplasmic membrane is composed of lipoprotein and phospholipids.
       2. Structure - The cytoplasmic membrane is a bilayer membrane.
       3. Functions - In bacteria, the cytoplasmic membrane is the site of enzymes of energy generation. It controls the exit and entrance of molecules by the mechanisms of diffusion and active transport.

VI. Cell Movement in Bacteria
    A. The major structures responsible for movement are flagella.
    B. Flagella have a relatively simple structure in bacteria. They consist of three parts: filament, hook, and basal body.

    C. Flagella push bacteria which swim through liquids.
       1. Bacteria move toward food sources and away from harmful materials, the process of chemotaxis.
       2. Chemotaxis is a primitive response and operates by determining the length of time the bacteria swim in one direction.

VII. Attachment of Bacterial Cells
    A. Pili are protein appendages of attachment.

VIII. Storage of Genetic Information in Bacteria
    A. The major structure in which the genetic information is stored is the chromosome.
    B. The bacterial chromosome is a double-stranded, circular DNA molecule.
       1. In prokaryotic cells only one or several identical chromosomes are present in each cell and it is not surrounded by a nuclear membrane.
       2. Basic proteins are not bound to the DNA.
    C. Plasmids are small DNA molecules that code for nonessential information and replicate independently of the chromosome.

IX. Synthesis of Proteins
    A. Ribosomes are structures on which proteins are synthesized.
       1. Ribosomes are composed of ribosomal proteins and ribosomal RNA molecules.

X. Storage Materials
    A. Bacteria often store a variety of materials that they later use as a source of nutrients. These include glycogen, volutin, and ß-hydroxybutyric acid.

XI. Endospores
    A. The endospore in a unique cell type that develops from actively multiplying cells (vegetative cells) when they face starvation.
    B. Endospores have unusual staining and are resistance properties.
    C. Many of the endospore-producing bacteria cause disease.

XII. Enclosure of Cytoplasm by Eukaryotic Cells
    A. The cytoplasmic (plasma) membrane is a lipid bilayer, enclosing the cytoplasm, similar in composition to the membrane in prokaryotic cells. Various glycoproteins are embedded in the membrane.
    B. Animal cells do not have a rigid cell wall.

XIII. Internal Membrane Structures in Eukaryotic Cells
    A. The endoplasmic reticulum (ER) encloses a large internal space and occurs in two forms. The rough ER has ribosomes for protein synthesis on its surface. The smooth ER lacks ribosomes and is involved in lipid metabolism.
    B. Golgi apparatus regulate the chemical modification and transport of molecules made in the endoplasmic reticulum.
    C. Peroxisomes contain enzymes which generate highly reactive toxic compounds in metabolism and also degrade these molecules.
    D. Lysosomes contain degradative enzymes such as nucleases and proteases.

XIV. Shape of Eukaryotic Cells
    A. The cytoskeleton consists of a network of protein filaments that give the cell its shape and that serve other roles as well. Actin filaments function in movement of cells. Microtubules are functional in cell movement by flagella and cilia. Microtubles also form the mitotic spindles that cause the movement of chromosomes during nuclear division.

XV.   Generation of Energy in Eukaryotic Cells
    A.   Mitochondria contain enzymes necessary for the generation of energy. The mitochondrial membranes function in ATP synthesis much like the cytoplasmic membrane of prokaryotic cells.
    B.   Mitochondria evolved from bacteria that were engulfed by cells. They contain circular DNA, ribosomes, and are the same size as bacteria.

XVI.   Storage of Genetic Information in Eukaryotic Cells
    A.   Genetic information is stored in double-stranded DNA associated with basic proteins (histones) to form chromatin. Segments of chromatin are the chromosomes of the cell. Each cell contains many chromosomes that carry different genes, the basic units of heredity. Each chromosome replicates and separates from its copy during cell division.
    B.   The chromosomes are enclosed by a bilayer membrane, the nuclear membrane, to form the nucleus. This is one feature that distinguishes eukaryotes from prokaryotes.

VOCABULARY: TERMS AND DEFINITIONS

compound microscope	**binary fission**	**basal body**
objective	**cell envelope**	**chemotaxis**
ocular	**peptidoglycan**	**plasmids**
resolution	**endotoxin**	**microtubles**
gram-positive	**halophilic**	**lysosomes**
acid-fast stain	**permeases**	**cytoskeleton**

1. Along with actin ———————————— are protein filaments that comprise a network or internal skeleton

   within the eukaryotic cell known as the ————————————— .

2. The ————————————————— is a microscope that has two sets of magnifying lenses. These

   are the ———————————— lens and the ———————————— lens.

3. Another name for membrane transport proteins that function to facilitate the movement of small molecules into and out

   of the cell is ———————————— .

4. ———————————— is the process of division by bacteria in which one cell divides to form two identical daughter cells.

5. The phenomenon of flagellated bacteria moving toward a nutrient source or away from harmful materials

   is called ————————————— .

6. ———————————— are the toxic lipid portions of the lipopolysaccharides of gram-negative bacteria.

7. Bacterial cells that retain the primary dye after decolorizing in the gram-staining procedure

   are said to be ————————————— .

8. ———————————— are membrane-bound organelles in eukaryotic cells that contain very powerful digestive enzymes.

9. The differential staining procedure used to identify the causative agents of tuberculosis and leprosy

   is the ————————————— .

10. _____ is a macromolecule found only in the cell wall of eubacteria.

11. Circular, double-stranded bits of DNA that are not part of the bacterial chromosome, are only a fraction of its size, and are not, except under certain conditions, essential for the survival of the cell are called _____ .

12. Bacterial flagella anchor into the cell wall and membrane by means of the _____ .

13. The capsule, the cell wall , and the cytoplasmic membrane are collectively referred to as the _____ .

14. Archaea found in the Great Salt Lake are described as "salt loving" cells or _____

15. The _____ of a microscope is a measure of its ability to separate small details.

Match the microscope from the list on the right with its description on the left.

_____16. Has optical devices to increase contrast between specimen and the background

_____17. Allows a three-dimensional view of an object's surface

_____18. Visualizes objects that emit light when light of a different wavelength strikes the object

_____19. Produces a map of the bumps and valleys of the atoms on the surface of a sample

_____20. Uses two beams of light to produce a three-dimensional appearance

_____21. Objects appear bright against a dark background

_____22. Electrons rather than light pass through the specimen

a. phase contrast microscope
b. fluorescence microscope
c. scanning tunneling microscope
d. darkfield microscope
e. transmission electron microscope
f. interference microscopes
g. scanning electron microscope

SELF-TEST OF READING MATERIAL

1. *Streptococcus mutans*, the major organism causing dental caries, attaches to the surface of teeth by a polysaccharide capsule. This capsule is synthesized only from
   a. diaminopimelic acid
   b. sucrose
   c. amino acids
   d. glucose
   e. lipid

2. The cell wall of gram-positive bacteria is composed primarily of
   a. chitin.
   b. cellulose.
   c. starch.
   d. protein.
   e. peptidoglycan.

3. Which of the following is essential for living cells?
   a. pili
   b. flagella
   c. cell wall
   d. cytoplasmic membrane
   e. capsule

4. Which of the following would contain peptidoglycan?
   1. ribosomes
   2. gram-positive cell walls
   3. pili

   4. cytoplasmic membrane
   5. flagella
   6. gram-negative cell walls

   a. 1 only
   b. 1,4
   c. 2,3,4

   d. 2 only
   e. 2,6

5. The process of cellular differentiation that produces an endospore from a vegetative cell is
   a. germination.
   b. common among the bacilli.
   c. sporogenesis.

   d. common among the cocci.
   e. Two of the above are correct.

6. In bacteria, sterols would be found only in the cell membranes of
   a. *Mycoplasma*.
   b. viruses.
   c. *Bacillus*.

   d. *Mycobacterium*.
   e. Sterols are never found in bacteria.

7. The group of bacteria that do not have cell walls are the
   a. Archaea.
   b. mycoplasma.
   c. eubacteria.

   d. mycobacteria.
   e. Both a and d are correct.

8. Cells that do not have a membrane separating their nuclear material from their cytoplasm are called
   a. heterozygous.
   b. prokaryotic.
   c. homozygous.

   d. eukaryotic.
   e. viruses.

9. Which of the following structures is unique to prokaryotic organisms?
   a. mitochondria.
   b. ribosomes.
   c. cell wall

   d. peptidoglycan cell wall
   e. cell membrane

10. Cell membranes are composed primarily of
    a. protein and polysaccharides.
    b. protein and phospholipid.
    c. phospholipid and polysaccharide.

    d. peptidoglycan in prokaryotes.
    e. Both b and d are correct.

11. Which of the following components are not found in the gram-positive cell wall
    a. lipopolysaccharides
    b. peptidoglycan
    c. teichoic acid

    d. N-acetyl muramic acid
    e. amino acids

12. Spherical bacteria arranged in a chain are known as
    a. spirochetes.
    b. bacilli.
    c. staphylococci.

    d. streptobacilli.
    e. streptococci.

13. Which of the following structures are known to protect bacteria from phagocytosis by white blood cells?
    a. flagella
    b. endospore
    c. capsule

    d. lister body
    e. ribosomes

14. Bacterial flagella and the flagella of eukaryotic cells have the same function but differ considerably in their structure.
    a. True
    b. False

15. Endotoxins are
    1. found in gram-positive cells
    2. lipopolysaccharides
    3. found in gram-negative cells
        a. 1,2,5
        b. 1,2
        c. 2,3
    4. lipoproteins
    5. found in *Mycoplasma*
        d. 3,4
        e. 3,4,5

16. Encapsulated bacteria
    a. are sometimes more virulent than their non-encapsulated counterparts.
    b. are "time-released" and easier to swallow than tablet bacteria.
    c. are more susceptible to phagocytic destruction.
    d. have more fun than non-encapsulated bacteria.
    e. both a and c are correct.

17. Bacterial flagella impart motility to the cell by
    a. undulating movement
    b. rotary movement
    c. gliding
    d. a and b
    e. b and c

REVIEW QUESTIONS

1. What are the two most important functions of a microscope?

2. List four essential life functions that all cells must have.

3. Why are penicillin and streptomycin active against bacteria but are not active against human cells?

4. List three ways in which endospores differ from vegetative cells.

NOTES

ANSWERS:
  Vocabulary: Terms and Definitions
    1. microtubules/cytoskeleton 2. compound microscope/objective/ocular 3. permeases 4. binary fission
    5. chemotaxis 6. endotoxins 7. gram-positive 8. lysosomes 9. acid-fast stain 10. peptidoglycan 11. plasmids
    12. basal body 13. cell envelope 14. halophilic 15. resolution 16. a 17. g 18. b 19. c 20. f 21. d 22. e
  Self-Test of Reading Material
    1. b 2.e 3. d 4. e 5. c 6. a 7. b 8. b 9. d 10. b 11. a 12. e 13. c 14. a 15. c 16. a 17. b

4 DYNAMICS OF BACTERIAL GROWTH

This chapter describes how microorganisms grow. Microorganisms are under the influence of the same factors that control the growth of any other organism. However, microorganisms do have some unique features in their growth patterns and their small size presents certain problems in measuring their growth. Bacteria multiply by binary fission in which one cell produces two identical daughter cells. When we describe the growth of bacteria we talk of the growth of populations rather than of individuals. This chapters explores the factors that affect the growth of microorganisms and describes how these factors often determine the habitat in which an organism grows.

KEY CONCEPTS

1. Bacterial growth is measured by an increase in cell numbers; growth of most other organisms is measured by an increase in their size.
2. All bacteria require a source of carbon to synthesize components of cytoplasm as well as a source of energy in order to grow. Different bacteria use different materials for their nutrients.
3. The growth requirements of bacteria may change permanently when they are taken from their natural environment and are grown in the laboratory.
4. Representatives of different groups of bacteria can multiply over a very broad range of environmental and nutritional conditions.

SUMMARY

I. Pure Culture Methods
   A. In nature, bacteria exist as mixed populations.
   B. Colonies represent the offspring of a single cell, and the streak-plate method is used to separate cells.
   C. The properties of agar make it ideal for growing bacteria on a solid surface.

II. Measurement of Cell Growth
   A. Bacterial Growth is defined in terms of population size, not by size of individual cells.
      1. Bacteria divide by binary fission.
      2. Number of bacteria increase exponentially.
   B. There are several methods for measuring bacterial growth.
      1. Scattering of light measures living and dead cells.
      2. Growth can be measured as an increase in cell number.
         a. The plate count measures the number of viable cells that can grow into a colony.
         b. Direct microscopic count is rapid but does not distinguish between living and dead bacteria.

III. Multiplication of Microorganisms - General Aspects
   A. Microorganisms grow and multiply throughout an enormous range of environmental conditions.
   B. An organism taken from its natural habitat often shows altered growth requirements when grown in the laboratory.

IV. Factors Influencing Microbial Growth
   A. The environmental factors that influence growth include temperature, oxygen, pH, and osmotic pressure.
      1. Temperature of Growth. Most bacteria grow within a temperature range of approximately 30°C
         a. Psychrophiles have an optimum growth temperature range between -5° and 20°C.
         b. Mesophiles have an optimum growth temperature range between 20° and 50°C.
         c. Thermophiles have an optimum growth temperature range between 50° and 80°C.
         d. Extreme thermophiles have an optimum growth temperature above 80°C.

2. Oxygen Requirements for Growth.
    a. Obligate (strict) aerobes have an absolute requirement for oxygen.
    b. Obligate (strict) anaerobes cannot multiply if oxygen is present because it kills the bacteria.
    c. Facultative anaerobes can use oxygen if it is available but can grow, although less well, in its absence.
    d. Microaerophilic organisms require small amounts of oxygen (2-10%) but higher concentrations are toxic.
    e. Aerotolerant organisms grow in the presence or absence of oxygen but they derive no benefit from the oxygen.
3. Aerobes require oxygen because they have metabolic pathways that require oxygen to convert energy in foodstuffs into forms useful to the cell.
    a. Bacteria that require oxygen have the means to detoxify certain forms of oxygen such as hydrogen peroxide.
    b. Detoxifying enzymes include catalase and superoxide dismutase.
4. Most bacteria grow best in a medium around neutrality (pH 7.0).
    a. Buffers are added to media to maintain a constant pH.
    b. Bacteria can maintain their internal contents at neutral pH even though the external pH is below or above this.
5. Osmotic pressure is determined by the concentration of dissolved substances in the medium in which bacteria are growing.
    a. Under normal conditions, the osmotic pressure is slightly higher on the inside than on the outside of the cell..
    b. Bacteria can compensate for pressure differences to a certain extent if the external osmotic pressure is high.

V. Nutritional Aspects of Bacterial Growth
  A. Cell multiplication requires a source of energy and raw materials for synthesis of cell components.
    1. Organisms demonstrate a spectrum of nutritional types based on their energy and carbon sources.
        a. Photoautotrophs use light (radiant) energy and CO_2 as a source of carbon.
        b. Photoheteroptophs use light (radiant) energy and organic compounds as a source of carbon.
        c. Chemoautotrophs use chemical energy and CO_2 as a source of carbon.
        d. Chemoheterotrophs use chemical energy and organic compounds as a source of carbon.
        e. Lithoheterotrophs use an inorganic source of chemical energy and organic compounds as a source of carbon.
    2. All organisms require a supply of nitrogen, sulfur, and phosphorous.
    3. Most organisms require small molecules, growth factors, that serve as subunits of biosynthetic macromolecules and cellular components.
        a. Some bacteria can synthesize all of their growth factors.
        b. Other bacteria, that lack the necessary biosynthetic enzymes, must have them provided.

VI. Cultivation of Bacteria in the Laboratory
  A. Bacteria are cultivated in the laboratory on defined or undefined media.
    1. Synthetic or defined media consist of chemically pure ingredients added in known amounts.
    2. Complex or undefined media consist of undefined nutrients such as ground meat, added in known amounts.

VII. Problems Associated with Growing Microorganisms in the Laboratory
  A. No single medium has been devised that will support the growth of all bacteria. Several factors may be responsible for this failure.
    1. Toxic compounds may be present and must be removed.
    2. Unusual growth factors may be needed.
    3. Some bacteria can only be grown in mixed culture, similar to their natural environment. One species provides the nutritional needs of the other.

B. Anaerobes require special cultivation methods.
1. An anaerobe jar may be used to remove all oxygen.
2. Chemicals may be added to combine with the oxygen.

VIII. Cultivation of Organisms that Occur as a Minor Part of the Microbial Population
A. Enrichment cultures enhance the growth of specific organisms.
B. Selective media preferentially inhibit the growth of certain organisms.

IX. Dynamics of Bacterial Population Growth - Four Well-recognized Phases
A. The lag phase of growth does not involve an increase in cell numbers, but is instead, a "tooling-up" phase.
B. During the exponential phase of growth, cell number increases exponentially.
C. Stationary phase is reached when the number of viable cells stops increasing.
D. Death phase is characterized by an exponential decrease in the number of viable cells.

X. Growth of a Bacterial Colony
A. Colony enlargement results from cells multiplying at the outside edges of the colony.

XI. Growth of Bacteria in Nature
A. In the laboratory, bacteria grow in a closed system; nutrients are not renewed and waste products are not removed.
B. An open system is the typical system found in nature. Nutrients are replenished and waste products are removed.
1. A chemostat in the laboratory simulates an open system in nature..
2. The doubling time as well as the population size of bacteria can be controlled in a chemostat.

VOCABULARY:TERMS AND DEFINITIONS

The following list contains new terms introduced in this chapter. Use these terms to fill-in the blanks of the sentences that follow and you will have a definition or description of each new term.

binary fission	agar	selective
generation time	buffers	chocolate agar
growth factors		

1. _____ are added to bacterial growth media to prevent the accumulation of acid by-products.

2. The time required for one cell to divide into two is the doubling time or _____ .

3. _____ is a polysaccharide extracted from marine algae and used as a solidifying agent in bacterial media.

4. A medium that contains substances to inhibit the growth of unwanted microorganisms is called

a _____ medium.

5. _____ are small organic molecules, other than energy sources, that must be provided for bacteria to grow.

6. A medium for the growth of nutritionally demanding bacteria that is enriched with blood and carefully heated to

release the growth factor hematin is called _____ .

7. When one cell divides and forms two identical daughter cells, the process is called _____ .

Match the category of organism from the list on the right with its description on the left. Some descriptions may have more than one answer and categories may be used more than once.

_____ 8. A bacterium with an optimal growth temperature of 37°C

_____ 9. A yeast that grows in the presence or absence of oxygen

_____ 10. A bacterium isolated from the intestines of an Arctic cod that is killed when exposed to the air

_____ 11. A mold that dies when incubated in an airtight container

_____ 12. An oxygen-dependent bacterium growing in a boiling hot springs in Yellowstone National Park

_____ 13. Usually must be grown in a candle jar because they cannot tolerate oxygen levels above 10%

_____ 14. *E. coli* that grows in our intestines

_____ 15. A bacterium with an optimal growth temperature of 62°C

a. psychrophile
b. obligate aerobe
c. thermophile
d. facultative anaerobe
e. mesophile
f. microaerophile
g. extreme thermophile
h. obligate anaerobe

SELF-TEST OF READING MATERIAL

1. Photoautotrophic metabolism involves
    1. the use of inorganic carbon dioxide as the sole source of carbon
    2. sunlight as the source of energy
    3. "rock eaters"
    4. the use of organic carbon compounds as a source of carbon
    5. energy derived from the oxidation of inorganic or organic molecules

    a. 1,2
    b. 1,2,3
    c. 3,4,5

    d. 4,5
    e. 3 only

2. Bacteria observed multiplying in the Dead Sea are most likely to be
    a. psychrophiles.
    b. anaerobic.
    c. mesophiles.

    d. halophiles.
    e. thermophiles.

3. Which of the following is essential for the growth of all bacteria?
    a. vitamins
    b. oxygen
    c. water

    d. glucose
    e. phospholipids

4. An organism that can grow either in the presence or in the absence of molecular oxygen is called a(n)
    a. obligate aerobe.
    b. facultative anaerobe.
    c. obligate anaerobe.

    d. facultative aerobe.
    e. halophile.

5. Cells make coenzymes from
    a. enzymes.
    b. carbohydrates.
    c. vitamins.

    d. proteins.
    e. apoenzymes.

6. Bacteria isolated from Antarctic marine fish are probably
   a. psychrophiles.
   b. mesophiles.
   c. halophiles.
   d. rockfordfiles.
   e. Both a and c are correct.

7. An 18 hour broth culture of *E. coli* contains

   a. *E. coli* cells that are all dividing exponentially.
   b. *E coli* cells that dividing at the same rate.
   c. mostly endospores.
   d. *E coli* cells that are all in stationary growth phase.
   e. *E coli* cells that are in all phases of growth.

Use the following curve to answer questions 8 through

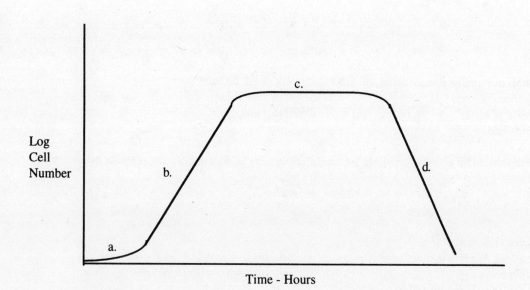

Log
Cell
Number

Time - Hours

8. During this phase of growth, bacteria in a population are adjusting to their medium.

9. During this phase of growth, bacteria are growing exponentially.

10. During this phase of growth, the bacterial population is doubling at a constant rate.

11. During this phase of growth, bacteria are dying exponentially.

12. During this phase of growth, cell division stops and cells do not die.

13. Bacteria with an optimum growth temperature around 10°C are called
    a. thermophiles.
    b. mesophiles.
    c. acidophiles
    d. psychopathic
    e. psychrophiles

14. The best definition of the generation time of a bacterial population is the
    a. time it takes for a bacterial population to double.
    b. time it takes for the lag phase.
    c. length of the exponential phase.
    d. time it takes for nuclear division.
    e. maximum rate of doubling.

15. A chemoheterotroph would use which of the following as a source of carbon?
    a. inorganic carbon
    b. carbon dioxide
    c. carbonate
    d. vitamins
    e. glucose

16. A bacterial culture began with 4 cells and ended with 128 cells. How many generations did the population go through?
    a. 64
    b. 32
    c. 6
    d. 5
    e. 4

17. Fungi generally grow best at a pH around
    a. 1
    b. 5
    c. 7
    d. 9
    e. 14

18. Bacterial broth cultures eventually stop growing and enter stationary phase because they
    a. deplete an essential nutrient.
    b. accumulate toxic products.
    c. become too crowded.
    d. Both a and b
    e. Both a and c

19. Which of the following could be added to a culture medium to neutralize acids and keep the pH near neutrality?
    a. buffers
    b. sugars
    c. agar
    d. heat
    e. water

20. A facultative anaerobic organism
    a. doesn't use oxygen, but is not killed by it.
    b. is killed by oxygen.
    c. grows in the presence or absence of oxygen.
    d. requires less oxygen than is present in air.
    e. prefers to grow without oxygen.

21. Agar is a good agent for solidifying liquid media because it
    a. is a good nutrient source.
    b. is a seaweed.
    c. liquefies at 40°C.
    d. solidifies around 100°C.
    e. is metabolized by very few bacteria.

22. Most bacteria grow best at pH
    a. 1
    b. 5
    c. 7
    d. 9
    e. 14

REVIEW QUESTIONS

1. Complete the following table.

NUTRITIONAL TYPE	SOURCE OF ENERGY	SOURCE OF CARBON
PHOTOAUTOTROPHS		
PHOTOHETEROTROPHS		
CHEMOAUTOTROPHS		
CHEMOHETEROTROPHS		

2. Describe the four stages of growth of a bacterial population in a broth medium.

3. Give an example of an environment from which you might isolate a) a thermophile, b) a mesophile, c) a psychrophile.

4. Distinguish between a selective and a differential medium.

NOTES

5 METABOLISM: THE GENERATION OF ENERGY AND SYNTHESIS OF SMALL MOLECULES

As an essential function of life, organisms take in nutrient molecules; degrade them to generate energy and smaller molecules; then, using the energy, they reassemble these small molecules to meet their own needs. This very basic process of life is known as metabolism. This chapter focuses on the catabolic portion of metabolism during which nutrient molecules are broken down to generate energy and small molecules. These products of catabolism become available for reassembly. Respiration and fermentation are the processes by which the cell accomplishes both the generation of energy and of small molecules. Many of these small molecules can be used for identification of bacterial species. The chapter ends with a look at some other ways in which microorganisms generate energy.

KEY CONCEPTS

1. Cells must perform two metabolic functions: (1) degrade foodstuffs to generate energy and small molecules, and (2) use the energy to convert the small molecules into marcomolecules and cell structures.
2. Foodstuffs are degraded by three major pathways, glycolysis, the TCA cycle, and the pentose-phosphate pathway.
3. The TCA cycle produces more energy than glycolysis for the same amount of foodstuff degraded.
4. Glycolysis and the TCA cycle have two functions: (1) to produce energy, and (2) to synthesize precursor metabolites that are converted into the subunits of macromolecules.
5. Bacteria as a group have a wide variety of enzymes and synthesize a large number of different products. Since these products are often unique for a species, identification of these products of metabolism can be used to identify bacterial species.
6. Some bacteria can oxidize a variety of inorganic compounds to gain energy.
7. Some bacteria are capable of photosynthesis and can use light as a source of energy.

SUMMARY

I. Overview of Metabolism
    A. Cells breakdown (catabolize) nutrient foodstuffs to generate energy and to produce precursor metabolites.
    B. The energy produced during catabolism is used to reassemble the precursor metabolites into macromolecules. This is called anabolism.

II. Enzymes - Chemical Kinetics and Mechanism of Action
    A. Enzymes rearrange atoms under the mild conditions of temperature and pressure at which cells can live.
    B. Enzymes speed up reactions by lowering their activation energy.
    C. The shape of the substrate must fit the shape of the enzyme that acts on it.
        1. An enzyme functions by first combining with its substrate through weak bonding forces.
        2. Next, the products of the reaction are released, leaving the enzymes unchanged.
    D. Environmental and chemical factors influence enzyme activity.
        1. Environmental factors include temperature, pH, and salt concentration.
            a. A $10^{\circ}C$ rise in temperature doubles the speed of enzymatic reactions.
            b. Most enzymes function best near a pH of 7.
            c. Salt concentration influences the shape of enzymes and therefore their activity.
        2. A variety of chemical compounds inhibits enzyme activity and the inhibition may be reversible or irreversible.
            a. When sulfa drugs compete with p-amino benzoic acid (PABA) for the active site of an enzyme used to synthesize the vitamin folic acid, the inhibition is reversible.
            b. An irreversible inhibitor may alter the amino acid sequence of the enzyme which in turn alters its shape and its activity.

III. Energy Metabolism
   A. Pathways of energy metabolism convert many low energy chemical bonds in a substrate to a few high energy bonds in the ATP molecule. This stored energy in ATP is now available to the cell for work.
   B. The two major pathways that cell have for generating energy are fermentation and respiration.
      1. In fermentation, an organic compound acts as the final hydrogen acceptor molecule.
      2. In respiration, an inorganic molecule acts as the final hydrogen acceptor molecule.
   C. Fermentation and respiration also generate precursor metabolites for the synthesis of cell components.
   D. Glycolysis is the most common pathway for the degradation (oxidation) of sugars.
      1. A glucose molecule metabolized (oxidized) through the glycolytic pathway produces 2 molecules of ATP, 2 molecules of pyruvic acid, and 2 molecules of NADH.
      2. Pyruvic acid is metabolized to different compounds depending on the species of bacterium and whether or not oxygen is present.
         a. If oxygen is not present, pyruvic acid is reduced to a variety of products by the NADH generated in glycolysis and the process is called fermentation. Examples of the products of fermentation include lactic acid from lactic acid fermentation, alcohol and carbon dioxide from alcoholic fermentation, and a variety of other compounds in mixed fermentations.
         b. If oxygen is present, pyruvic acid can be further oxidized to carbon dioxide and water.

IV. Respiration
   A. During respiration, cells oxidize pyruvic acid to carbon dioxide (CO_2) and water (H_2O) and trap the energy from pyruvic acid in ATP.
      1. In the first step, hydrogens are removed from the pyruvic acid molecule (oxidation) along with a carbon atom. The remaining two carbon fragment attaches to coenzyme A and is further oxidized in the tricarboxylic acid cycle (TCA cycle).
      2. In the TCA cycle, coenzyme A is regenerated, hydrogens are removed (oxidation) and transferred to NAD, and the two carbon fragment is released as carbon dioxide (CO_2).
      3. The electrons from the hydrogen atoms that have been removed since glycolysis are transported through the electron transport chain. This transfer of electrons eventually generates energy to synthesize ATP from ADP and inorganic phosphate. This is known as oxidative phosphorylation.
   B. About 20 times more ATP can be synthesized from respiration than from fermentation.
      1. Fermentation produces a net gain of 2 ATP for each molecule of glucose oxidized.
      2. Respiration produces up to a net gain of 38 ATP for each molecule of glucose oxidized.

V. Respiration Under Anaerobic Conditions
   A. Some bacteria can utilize inorganic compounds other than oxygen as final electron acceptors. Nitrate, sulfate, and carbon dioxide are the most common acceptors in anaerobic respiration.
   B. Less ATP is synthesized during anaerobic respiration compared to aerobic respiration.

VI. Metabolism of Compounds Other than Glucose
   A. Many bacteria have alternative pathways that utilize compounds other than glucose for energy metabolism.
   B. Most organic compounds that are catabolized, including amino acids, lipids, and carbohydrates, are converted into compounds that are part of either glycolysis or the TCA cycle.

VII. Other Pathways of Glucose Degradation
   A. Entner-Doudoroff Pathway
   B. The pentose-phosphate pathway provides reducing power for biosynthesis as well as certain precursor metabolites.

VIII. Oxidation of Inorganic Compounds
      1. Some bacteria can oxidize a variety of inorganic compounds such as ammonia, hydrogen sulfide, and hydrogen gas to generate the energy needed to synthesize ATP.

IX. Biosynthetic Metabolism
   A. Degradative and biosynthetic pathways may be part of the same pathway.
      1. Degradative pathways generate the molecules that serve as the starting materials (precursor metabolites) for the synthesis of cell components.

B. Many amino acids are synthesized from the same precursor metabolites. A very important reaction in biosynthetic metabolism is the incorporation of nitrogen into organic compounds to form amino acids.

    1. Ammonium (NH_4^+) is added to α–ketoglutaric acid to form glutamic acid. This reaction is reversible and connects anabolism with catabolism.

    2. The ammonium (NH_4^+) can be transferred to a variety of other α–keto acids to form other amino acids.

C. Purines and pyrimidines are synthesized stepwise along different pathways.

X. Bacterial Photosynthesis

A. Radiant energy (light) absorbed by chlorophyll provides the energy to reduce carbon dioxide to organic compounds.

B. Compounds, such as hydrogen sulfide (H_2S), serve in place of water (H_2O) as the reducing agent for photosynthetic bacteria, with the exception of the cyanobacteria, which do use water.

    1. If oxygen is released, then photosynthesis is called oxygenic.

    2. If something other than oxygen is released, then it is called anoxygenic.

C. Photosynthesis requires light for energy generation, but biosynthetic reactions do not.

    1. "Light" reactions occur only in photosynthetic organisms that gain energy from light. Electrons are transferred along an electron transport chain and ATP is synthesized.

    2. The biosynthesis of carbohydrates does not require light, only the ATP from the "light" reactions.

    3. In these "dark" reactions, carbon dioxide is added to a 5-carbon compound ribulose 1,5-biphosphate to form a metabolite in the glycolytic pathway.

VOCABULARY:TERMS AND DEFINITIONS

The following list contains new terms introduced in this chapter. Use these terms to fill-in the blanks of the sentences that follow and you will have a definition or description of each new term.

cytochromes/quinones	**glycolysis**	**oxidation**
competitive	**precursor molecules**	**coenzymes**
reduction	**activation**	**active/catalytic**

1. _____ are organic molecules that are converted into products by enzymatic reactions to become subunits of macromolecules.

2. Energy that must be supplied to a chemical reaction in order to get it going is called the

    energy of _____ .

3. _____ is the addition of electrons or hydrogen atoms to a molecule.

4. The process in cells that involves the oxidation of glucose to pyruvic acid with the production of 2 ATP

    is known as _____ .

5. Sulfa drugs act on bacterial cells by _____ inhibition.

6. _____ and _____ are electron carriers in the electron transport system.

7. The removal of electrons or hydrogen atoms from a molecule is called _____ .

8. _____ are small nonprotein molecules that function in the transfer of atoms from one molecule to another.

9. The point at which a substrate attaches to its specific enzyme is called the _____ or the

    _____ site.

SELF-TEST OF READING MATERIAL

1. During aerobic respiration, the final hydrogen acceptor molecule is
    a. an organic compound.
    b. pyruvic acid.
    c. oxygen.
    d. cytochrome.
    e. glucose.

2. Reduction is the
    a. addition of protons.
    b. removal of electrons.
    c. addition of oxygen.
    d. removal of electrons or hydrogen atoms.
    e. addition of electrons or hydrogen atoms.

3. The amount of ATP synthesized during aerobic respiration is
    a. much less than that from fermentation.
    b. greater than that from fermentation.
    c. about the same as that from fermentation.

4. The catalytic activity of an enzyme is determined by the
    a. substrate.
    b. presence of ions.
    c. 3-dimensional shape of the enzyme.
    d. presence of vitamins.
    e. hydrolysis of ATP.

5. What is the net gain of ATP from the oxidation of glucose to carbon dioxide and alcohol?
    a. 2
    b. 4
    c. 34
    d. 38
    e. 40

6. Which of the following is not an enzyme?
    a. dehydrogenase
    b. cellulase
    c. coenzyme A
    d. lactase
    e. hexokinase

7. The ATP produced during glycolysis is the result of
    a. substrate phosphorylation.
    b. oxidative reduction.
    c. substrate reduction.
    d. oxidative phosphorylation.
    e. photophosphorylation.

8. Bacteria need energy for
    a. synthesizing macromolecules.
    b. cell division.
    c. motility.
    d. catabolism.
    e. All of the above processes require energy.

9. The synthesis of ATP from ADP and inorganic phosphate in which the source of energy is from the transfer of electrons is called
    a. substrate reduction.
    b. reductive phosphorylation.
    c. substrate oxidation.
    d. photooxidation.
    e. oxidative phosphorylation.

10. The molecules on which enzymes work are called
    a. substrates.
    b. inhibitors.
    c. active molecules.
    d. coenzymes.
    e. ATP.

11. Which of the following competitively inhibits enzyme activity in the synthesis of folic acid?
    a. sulfa drugs
    b. vitamins
    c. penicillin
    d. PABA
    e. folic acid

12. Which of the following would not be a product of fermentation?
    a. acetic acid
    b. ethyl alcohol.
    c. carbon dioxide
    d. pyruvic acid
    e. lactic acid.

13. Approximately what percent of the energy in a glucose molecule is converted to ATP during aerobic respiration?
    a. 5%
    b. 11%
    c. 50%
    d. 89%
    e. 95%

14. Catalase is the enzyme involved in
    a. substrate phosphorylation.
    b. the synthesis of hydrogen peroxide.
    c. the breakdown of acetoin.
    d. the synthesis of acetoin.
    e. the breakdown of hydrogen peroxide.

15. Anaerobic respiration might use which of the following electron acceptors?
    a. glucose
    b. pyruvate
    c. oxygen
    d. cytochrome
    e. nitrate

16. Oxidation is the
    a. addition of protons.
    b. addition of neutrons.
    c. removal of protons.
    d. removal of electrons or hydrogen atoms.
    e. addition of electrons or hydrogen atoms.

17. Hydrogens removed during the oxidation of glucose are usually transferred to
    a. INS.
    b. ATP.
    c. NAD.
    d. DNA.
    e. ADP.

18. During fermentation, glucose is oxidized to
    a. sugar.
    b. water.
    c. a variety of products depending on the organism.
    d. oxygen.
    e. fermentation is reduction and not oxidation.

19. Enzymes work by supplying the activation energy required for a chemical reaction to take place.
    a. True
    b. False

20. In eukaryotic cells, glycolysis takes place in the mitochondrian.
    a. True
    b. False

REVIEW QUESTIONS

1. Distinguish among substrate phosphorylation, oxidative phosphorylation, and photophosphorylation.

2. Complete the following table by indicating the number of molecules of ATP, NADH, and FADH produced from one glucose molecule.

	ATP	NADH	FADH
GLYCOLYSIS			
PYRUVIC ACID TO ACETYL CoA			
TCA CYCLE			

3. Summarize the reactions and products of glycolysis in a single sentence.

4. How do the end products of photosynthesis in cyanobacteria and green plants differ?

ANSWERS:
    Vocabulary:Terms and Definitions
        1. precursor metabolites 2. activation 3. reduction 4. glycolysis 5. competitive 6. cytochromes/quinones
        7. oxidation 8. coenzymes 9. active/catalytic
    Self-Test of Reading Material
        1. c 2. e 3. b 4. c 5.a 6. c 7. a 8. e 9. e 10. a 11. a 12. d 13. c 14. e 15. e 16. d 17. c 18. c 19. b
        20. b

6 INFORMATIONAL MACROMOLECULES: FUNCTION, SYNTHESIS AND REGULATION OF ACTIVITY

In order to reproduce and to carry on the many activities associated with life, cells must have a plan or blueprint to follow. These instructions are carried in the form of a macromolecule called DNA. DNA is an amazing molecule not only because it contains the codes for every life function, but also because it reproduces itself and rarely makes a mistake. In addition to replication, DNA also directs the synthesis of another macromolecule, RNA, that serves as a messenger in the process of protein synthesis. The code for life is actually a code for the synthesis of proteins. Cells monitor their environment and synthesize only those proteins that are necessary at that particular moment. Cells have a means of control or regulation of their protein synthesis.

KEY CONCEPTS

1. The general rule for the flow of information in cells is:

	codes for		codes for	
DNA	———————	RNA	———————	protein
	the synthesis of		the synthesis of	

2. In the transfer of information, a sequence of nucleotides in DNA determines a sequence of amino acids in proteins.
3. DNA has two functions: to store genetic information and to reproduce itself.
4. The sequence of nucleotides in one strand of DNA determines the sequence in the other strand.
5. The enzymes of bacteria may change with changes in the environment, because it turns on and off gene functions.
6. The synthesis of many small molecules, such as amino acids, is controlled by regulating the activity as well as the number of enzyme molecules involved in the synthesis of the amino acid.
7. Bacteria synthesize only the amounts of each cell component that they need in order to multiply most rapidly.

SUMMARY

I. Chemistry of DNA and RNA
   A. DNA occurs as a double-stranded helix held together by hydrogen bonds between adenine and thymine, and between guanine and cytosine.
   B. The two strands run in opposite direction, one going fromthe top to the bottom; the other from the bottom to the top.
   C. RNA is usually single-stranded and shorter than DNA.

II. Information Storage and Transfer
   A. Information is stored in DNA and RNA in the sequence of their nucleotide subunits.
   B. Information in DNA can be transferred by two different processes.
      1. The sequence of nucleotides in DNA can be transferred to mRNA, which determines the sequence of amino acids in proteins.
      2. Replication of DNA begins at a specific site on the chromosome and proceeds by the sequential addition of nucleotides by DNA polymerase. Two chromosomes identical to the original are formed.

III. The Expression of Genes
   A. A bacterial chromosome consists of two major regions. One codes for proteins (coding region); the other determines whether the first region will be expressed (regulatory region).
   B. The sequence of three nucleotides in mRNA determines a specific amino acid, a triplet code.

    C.   Gene transcription is the process of making an RNA copy of the genetic information in DNA.
        1.   The three steps of transcription are:
            a.   Initiation - The sigma factor of RNA polymerase binds to the promoter region in the regulatory region of the gene.
            b.   Elongation - RNA polymerase moves along one of the strands of DNA (template strand) synthesizing a complementary single strand of mRNA.
            c.   Termination - RNA polymerase comes to a stop signal at the end of a gene and is released.
    D.   The second step in gene expression is the translation of mRNA into protein.
        1.   The sets of three nucleotides (codons) that code for each amino acid have been identified. Most amino acids have several codons.
        2.   Translation involves three steps.
            a.   Initiation - The ribosome binds to a region near the beginning of the mRNA to start translation.
            b.   Elongation - A charged tRNA molecule attaches to its complementary codon of the mRNA positioned at a particular site on the ribosome. The ribosome moves along the mRNA three nucleotides at a time, so that another charged tRNA bonds to the next complementary codon. The amino acids are joined together in a peptide bond.
            c.   Termination - When the ribosome reaches a stop codon in the mRNA, the protein dissociates from the mRNA, and protein synthesis stops.

IV.   Differences Between Prokaryotic and Eukaryotic Gene Transcription/Translation
    A.   There are a number of differences between prokaryotic and eukaryotic gene transcription/translation.
        1.   In prokaryotic cells, transcription and translation occur almost simultaneously; in eukaryotes, mRNA is transported out of the nucleus before translation begins.
        2.   In eukaryotes, most mRNA molecules are modified before they are translated. Blocks of nucleotides are cut out, a situation very rare in prokaryotes.

V.   The Environment and Control Systems
    A.   Bacteria may be subjected to rapidly changing nutrients in the environment. They have the means to shut off or to activate pathways of biosynthesis and degradation so as to synthesize only the materials they need in any particular environment.

VI.   Mechanisms of Control of Biosynthetic Pathways
    A.   End product repression controls the start of gene transcription.
        1.   The end product binds to the repressor and changes its shape so that it can then bind to the operator region of the gene. This prevents the RNA from binding to the promoter.
    B.   End products control the activity of the first enzyme of the biosynthetic pathway (allosteric control).
        1.   The end product combines with the first enzyme of the pathway and changes its shape so it can no longer catalyze its reaction. This inhibition of the first enzyme shuts down the entire pathway.

VII.   Mechanisms of Control of Degradative Pathways
    A.   Metabolites that can be degraded activate the genes responsible for their degradation.
        1.   The compound to be degraded turns on the pathway that can degrade it. The compound binds to the repressor protein and inactivates it, thereby allowing RNA polymerase to initiate transcription.
    B.   Catabolite repression ensures efficient use of available foodstuffs.
        1.   A catabolite repressor can reduce the level of cyclic AMP required for induction of certain pathways, so the catabolite repressor is metabolized first.
    C.   Some very important enzymes are synthesized at the same level independent of the medium. These are called constitutive enzymes.

VOCABULARY:TERMS AND DEFINITIONS

The following list contains new terms introduced in this chapter. Use these terms to fill-in the blanks of the sentences that follow and you will have a definition or description of each new term.

template	semiconservative	codon/anticodon
coding region	fork	nucleotides
transcription	end product/feed back	antiparallel
operon	regulator region	inducers
complementary		

1. The _____ of a gene determines the sequence of amino acids in a polypeptide chain.

2. The subunits or building blocks for both DNA and RNA are called _____ .

3. The name _____ is given to a set of genes which are linked together and which are transcribed as a unit.

4. The complementary strands in a DNA molecule are arranged in opposite directions and are said

    to be _____ .

5. The name _____ is given to a set of three nucleotides on mRNA that code for a particular amino

    acid. The name _____ is given to its complementary set of nucleotides on tRNA.

6. DNA replication is described as _____ because each new molecule contains one new DNA strand and one of the original DNA strands.

7. One strand of a DNA molecule is said to be _____ to the other strand.

8. RNA synthesis in which the genetic code is transferred from DNA to mRNA

    is called _____ .

9. The sequence of nucleotides in DNA serves as a _____ or pattern which is copied into A complementary strand of either DNA or mRNA.

10. The _____ of a gene does not code for amino acids, but rather contains nucleotide sequences where molecules bind and control the expression of the gene.

11. The point at which nucleotides are added during DNA replication is called the

    replication _____ .

12. The two main methods by which biosynthetic pathways are controlled are _____ repression

    and _____ inhibition.

13. The metabolites that activate gene transcription are called _____ .

SELF-TEST OF READING MATERIAL

1. The best definition of a gene is
   a. a sequence of nucleotides in DNA that codes for a functional product.
   b. a segment of DNA.
   c. three nucleotides that code for an amino acid.
   d. a sequence of nucleotides in RNA that codes for a functional product.
   e. a segment of RNA.

2. Which of the following is a product of transcription?
   a. a new strand of DNA.
   b. protein
   c. mRNA
   d. polypeptide
   e. All of the above are products of transcription.

3. If a segment of one strand of a DNA molecule had the nucleotide sequence AGTTACGGTAAT, what would the sequence on the complimentary strand be?
   a. UCAAUGCCAUUA
   b. ATTACCGTAACT
   c. TCAATGCCATTA
   d. TGCAATGGATCG
   e. AGTTACGGTAAT

4. Which of the following RNA's transports the amino acids to the ribosomes during protein synthesis?
   a. cRNA
   b. rRNA
   c. mRNA
   d. tRNA
   e. fRNA

5. In a DNA molecule, the following base pairing is normally observed
   1. A-G
   2. C-G
   3. A-T
   4. A-U
   5. U-C
   a. 1,2
   b. 2,3
   c. 2,4
   d. 3,5
   e. 1,5

6. The backbone of the strands of a DNA molecule is made of alternating units of
   a. deoxyribose and purine.
   b. ribose and purine.
   c. deoxyribose and pyrimidine.
   d. ribose and phosphate.
   e. deoxyribose and phosphate.

7. The chemical composition of DNA includes which of the following?
   1. uracil
   2. ribose
   3. thymine
   4. deoxyribose
   5. one strand
   6. two strands

   a. 1,4,5
   b. 2,3,6
   c. 1,2,6
   d. 2,3,5
   e. 3,4,6

8. In prokaryotic cells, transcription and translation both occur in the cytoplasm while in eukaryotic cells transcription occurs in the nucleus and translation occurs in the cytoplasm.
   a. True
   b. False

9. Translation is also known as
   a. DNA synthesis.
   b. protein hydrolysis.
   c. reverse transcription.
   d. RNA replication
   e. protein synthesis.

10. The site of protein synthesis in prokaryotic cells is the
    a. mitochondrion
    b. LPS.
    c. ribosome.
    d. cell membrane.
    e. nucleoid.

11. The metabolites that activate gene transcription are called
    a. repressors.
    b. operators.
    c. inducers.
    d. synthesizers.
    e. promoters.

12. Feedback inhibition involves the inhibition of only the first enzyme of the biosynthetic pathway while end product repression prevents the synthesis of all of the enzymes of the pathway.
    a. True
    b. False

13. Anticodons are found on which of the following types of nucleic acids?
    a. DNA
    b. tRNA
    c. cDNA
    d. mRNA
    e. rRNA

14. Initiation, elongation and termination are steps involved in which of the following processes?
    a. transcription
    b. translation
    c. replication
    d. a and b are both correct.
    e. a and c are both correct.

15. Enzymes not subject to regulation by induction or repression are called
    a. repressor enzymes.
    b. constitutive enzymes.
    c. penicillinase.
    d. promoter enzymes.
    e. inducible enzymes.

REVIEW QUESTIONS

1. Diagram the flow of information during DNA replication, and RNA and protein synthesis.

2. Compare the functions of the three types of RNA.

3. Distinguish between feedback inhibition and end product repression.

4. Complete the following table describing three major differences in transcription and translation in prokaryotes and eukaryotes.

PROKARYOTES	EUKARYOTES
1. Transcription and translation occur together	1.
2.	2. mRNA is processed before translation
3.	3. mRNA often transcribed as polygenic mRNA

ANSWERS:
  Vocabulary:Terms and Definitions
    1. coding region 2. nucleotides 3. operon 4. antiparallel 5. codon/anticodon 6. semiconservative
    7. complementary 8. transcription 9. template 10. regulator region 11. fork 12. end product/ feed back
    13. inducers
  Self-Test of Reading Material
    1. a 2. c 3. c 4. d 5. b 6. e 7. e 8. a 9. e 10. c 11. c 12. a 13. b 14. d 15. b

7 BACTERIAL GENETICS

Bacterial genetics is probably the fastest growing area of microbiology today. Elucidation of the chemical structure of DNA has quickly led to an understanding of how cells function at a molecular level. This chapter describes the structure of DNA as well as how changes in that structure produces mutations. These mutations are spontaneous but can increase in frequency in the presence of mutagens. Bacteria are not considered sexually reproducing organisms, but there are processes by which some bacteria can undergo genetic recombination. These processes are used to better understand the genetics of bacteria.

KEY CONCEPTS

1. Many chemicals as well as ultraviolet light causes changes in DNA that are passed on to future generations.
2. In bacteria, small pieces of the chromosome can be transferred from one cell to another.
3. Genes can move from one location in DNA to other locations in the DNA of the same cell.
4. Plasmids confer a large number of different, but dispensable properties on bacteria, and are readily transferred to other bacteria, often unrelated to each other.
5. Bacteria are able to protect themselves from foreign DNA entering the cell.

SUMMARY

I. Source of Diversity in Microorganisms
   A. Two major sources account for diversity
      1. Variation in sequences of nucleotides that exist in DNA.
      2. The environment regulates gene function and enzyme activity.

II. Alterations in Nucleotide Sequence in DNA
   A. Mutations can occur by substitution of one nucleotide for another.
   B. Mutations can occur by the removal or addition of nucleotides.
   C. Mutagens increase the frequency of mutations.

III. Mutagenesis is the Process by which a Mutation is Produced
   A. Spontaneous mutations occur in the absence of any known mutagen. These mutations are the same as those caused by mutagens.
   B. Three types of mutagens are known: chemical agents, transposable elements, and radiation.
   C. Chemical mutagens often act by altering the hydrogen-bonding properties of the bases.
      1. Chemical mutagens include intercalating agents, base analogs, and alkylating agents.
      2. Base analogs are incorporated into DNA in place of the natural base.
      3. Intercalating agents result in the addition of nucleotides.

IV. Transposable Elements
   A. Transposable elements are segments of DNA that move from one site in a DNA molecule to another. They insert into genes and disrupt their function.
   B. Insertion sequence is the simplest transposable element. It is composed of two inverted sequences flanking an enzyme required for transposition.
   C. More complex transposable elements often code for antibiotic resistance.

V. Radiation
   A. Ultraviolet light and X-rays are mutagenic.
   B. Ultraviolet light causes the formation of thymine dimers in the DNA molecule.
   C. X-rays cause breaks in the DNA molecule.

VI.   Repair of Damaged DNA
    A.   UV damage can be overcome by an inducible enzyme system of DNA replication that bypasses the damaged DNA, but makes many mistakes. This is the SOS system.
    B.   Visible light activates an enzyme that breaks the covalent bonds forming thymine dimers. This is light repair.
    C.   Bacteria can excise damaged DNA and replace it with undamaged DNA. This is excision repair or dark repair.

VII. Rates of Mutation
    A.   Genes mutate independently of one another and mutations are stable.
    B.   Mutations are expressed rapidly in prokaryotic cells because they have only one or two identical chromosomes.

VIII. Mutant Selection
    A.   Direct Selection requires that the mutant grow on a solid medium on which the parent cannot.
    B.   Indirect selection must be used to isolate mutants that cannot grow on a medium on which the parent can grow.
        1.   Replica plating involves transferring all of the colonies on one plate to another plate simultaneously.
        2.   Penicillin enrichment increases the proportion of mutants in a population. It is based on the observation that penicillin kills only growing cells.

IX.   Conditional Lethal Mutants
    A.   Conditional lethal mutants result from mutations in genes that would ordinarily be lethal.
    B.   These conditional lethal mutations are usually in genes that are involved in macromolecular synthesis.

X.   Commercial Applications of Mutants
    A.   Chemicals can be tested for their cancer-causing ability by their mutagenic activity. The Ames test measures the frequency of reversion of his⁻ to his⁺ cells.

XI.   Mechanism of Gene Transfer
    A.   Genes can be transferred between bacteria by three different mechanisms.
    B.   Only short segments of DNA can be transferred.
    C.   Recombination between donor and recipient DNA occurs by a breakage and reunion mechanism.
        1.   DNA-mediated transformation involves the transfer of naked DNA. Recipient cells must be able to take up DNA through their envelope - natural competence. Cells that are not naturally competent can be made artificially competent by punching holes in their envelopes with an electric current.
        2.   Transduction involves viral transfer of DNA.
        3.   Conjugation requires cell-to-cell contact. Usually only plasmids are transferred from F⁺ to F⁻ cells. Hfr cells are formed if the F⁺ plasmid integrates into the chromosome. Hfr cells can transfer chromosomal DNA. F' donor cells result when the F⁺ plasmid integrated into the chromosome is excised and carries a piece of chromosomal DNA with the plasmid.

XII. Plasmids
    A.   R-Factor plasmids confer resistance to a large number of antimicrobial agents. They can be transferred to related and unrelated bacteria by conjugation.

XIII. Gene Transfer Within the Same Bacterium
    A.   Transposable elements are important for the transfer of genes to unrelated bacteria because they can move from a chromosomal site to a plasmid which can then be transferred by conjugation.

XIV. Restriction and Modification of DNA
    A.   DNA entering unrelated bacteria is recognized as being foreign and is degraded by deoxyribonucleases termed restriction enzymes because they cleave the DNA at restricted or specific sites.
    B.   Modification enzymes confer resistance to restriction by altering the nucleotides in the DNA at the sites the restriction enzymes cleave the DNA.

XV.   Importance of Gene Transfer to Bacteria
    A.   Gene transfer allows microorganisms to survive changing environments by providing recipient cells with whole new sets of genes.

XVI.   Importance of Gene Transfer to the Microbial Geneticist
    A.   Gene mapping - Genes that are close together will be transferred together by conjugation, transformation, and transduction. Analysis of these transfers allows for the calculation of relative distances between genes.
    B.   Genetic Engineering - Gene transfer is the method that genetic engineers use to introduce DNA into bacteria.

VOCABULARY:TERMS AND DEFINITIONS

The following list contains new terms introduced in this chapter. Use these terms to fill-in the blanks of the sentences that follow and you will have a definition or description of each new term.

restriction enzymes	**sex pilus**	**mutagens**
episome	**haploid**	**frame shift**
bacteriophage	**auxotrophs/prototrophs**	**genotype/phenotype**
Ames test	**base analogs**	**mutation**
reversion	**transposable elements/transposons**	

1. The genetic constitution of an organism is its ———————————— while the characteristics displayed by an organism in a given environment comprise the ———————————— of the organism.

2. Bacteria are called ———————————— because they contain one or sometimes several identical chromosomes, but never pairs of chromosomes.

3. ———————————— protect bacterial cells from foreign DNA entering them.

4. The deletion or addition of one or more nucleotides in a DNA molecule causes a misreading of the codons and is termed a ———————————— mutation.

5. ———————————— is a simple, inexpensive, and quick microbiological test for potential carcinogens.

6. The ———————————— is an appendage on a donor cell that attaches to a recipient cell during conjugation.

7. Compounds that resemble the chemical structure of nitrogen bases closely enough that they are incorporated into DNA and result in mutations are know as ———————————— .

8. Bacterial viruses are also called ———————————— .

9. Chemical or physical agents that increase the frequency of mutations are called ———————————— .

10. An ———————————— is a bacterium that requires a growth factor in order to grow while a ———————————— does not require a growth factor.

11. A change in the nucleotide sequence of DNA which results in a recognizable change in the organism is called a ———————————— .

12. DNA that may exist either extrachromosomally or as part of the chromosome is termed an ——————————— .

13. ———————————— is the process by which a cell phenotype changes back to its original state through a mutation.

14. ———————————————— are special segments of DNA that can move from one site in a DNA molecule to another. Another name for these special segments is ———————————————— .

SELF-TEST OF READING MATERIAL

1. The genome of an organism includes genes from
    a. chromosomes.
    b. nucleotides.
    c. plasmids.
    d. a and b
    e. a and c

2. Genotypic changes are rare and usually involve only a few cells in a population while phenotypic changes will involve almost all of the cells in a population.
    a. True
    b. False

3. In the Ames test for mutagens, extract of rat liver is added to the test because
    a. it is a good source of nutrients.
    b. the extract can convert test compounds to active carcinogens.
    c. it contains mRNA necessary for the test.
    d. it neutralizes toxic materials in the test.
    e. it contains DNA necessary for the test.

4. Resistance to many antibiotics is carried on
    a. R-factor plasmids.
    b. sex pili.
    c. enzymes.
    d. chromosomes.
    e. A factor plasmids.

5. Ultraviolet light is a selectively absorbed in the cell by
    a. enzymes.
    b. DNA.
    c. cell membranes.
    d. proteins.
    e. RNA.

6. Which of the following chemical mutagens will result in the addition of nucleotides to the DNA?
    a. base analogs
    b. alkylating agents
    c. intercalating agents
    d. nitrous acid
    e. All of the above agents.

7. Mutations are expressed more frequently in bacteria because they
    a. divide often.
    b. are competent.
    c. divide infrequently.
    d. are diploid.
    e. are haploid.

8. The technique of replica plating involves transferring all of the colonies on one plate simultaneously to another plate.
    a. True
    b. False

9. The segments of DNA transferred between cells are
    a. large.
    b. single stranded.
    c. small.
    d. circular.
    e. not functional.

10. Thymine dimer formation is the result of exposing DNA to
    a. ultraviolet light.
    b. base analogs.
    c. transposons.
    d. X-rays.
    e. intercalating agents.

11. The process of DNA repair in which enzymes cut out a damaged section of DNA and other enzymes then repair the resulting break is known as
    a. excision repair.
    b. SOS repair.
    c. photoreactivation.
    d. light repair.
    e. DNA cannot be repaired.

12. When bacteria develop resistance to an antibiotic, the antibiotic does not cause the mutation but only selects for a pre-existing mutation.
    a. True
    b. False

13. An organism which requires a growth factor in order to grow is called a(n)
    a. prototroph.
    b. wild type.
    c. autotroph.
    d. auxotroph.
    e. heterotroph.

14. The Ames test for chemical mutagens
    1. is less expensive than animal testing.
    2. uses the mutation of a his+ cell.
    3. is faster than animal testing.
    4. is more expensive than animal testing.
    5. uses the reversion of a his- mutant.
    6. takes longer than animal testing.

    a. 1,2,3
    b. 1,3,5
    c. 1,2,6
    d. 4,5,6
    e. 2,4,6

15. Conjugation differs from transformation and transduction in that it requires contact between the donor and the recipient.
    a. True
    b. False

REVIEW QUESTIONS

1. Compare light and dark repair of DNA.

2. Distinguish between genotype and phenotype.

3. Describe three ways in which mutations may occur.

4. What is the relationship between mutagens and carcinogens.

5. List three advantages of the Ames teat compared to animal testing.

8 MICROBIOLOGY AND BIOTECHNOLOGY

Terms such as gene cloning, recombinant DNA, and genetic engineering are becoming commonplace in our language and our lives. Hardly a week goes by without one of these terms appearing in a news magazine or on the evening news broadcast. This chapter introduces you to the application of microbial genetics to areas of biotechnology. The methods of gene cloning, genetic engineering, and the polymerase chain reaction, which did not exist 20 years ago, are explored with examples of how these new technologies are important in today's society.

KEY CONCEPTS

1. DNA is a macromolecule that is easily studied and manipulated in a test tube.
2. The discovery of restriction enzymes opened the door to gene cloning and genetic engineering.
3. Gene cloning can provide large amounts of specific DNA for study as well as large quantities of products coded by the cloned DNA.
4. In the laboratory, DNA from one organism can be introduced into other organisms where it will replicate and be expressed.
5. The polymerase chain reaction is among the most important technical advances made in molecular biology in the past decade.

SUMMARY

I. Applications of Biotechnology in Industry
   A. Medically important proteins such as insulin, human growth factor, and blood clotting factors can be produced by *Escherichia coli* and yeast.
   B. Safer vaccines can be made using genetically engineered microorganisms.
   C. Bioremediation involves the use of microorganisms to degrade harmful chemicals.
   D. Plants can be genetically engineered to give them new properties.
   E. Toxins from bacteria can serve as effective biocontrol agents

II. Cloning
   A. Cloning is removing a DNA fragment from an organism and introducing it into another such as *Escherichia coli* or yeast. Cloning involves a number of steps.
   B. The cells are broken up to release all of their DNA.
   C. Restriction enzymes and ligase are the tools of molecular biology.
   D. Vectors act as a self-replicating carrier of foreign DNA.
   E. Ligating the vector and the insert together creates a recombinant plasmid.
   F. The recombinant plasmid must be introduced into a new host cell.
   G. Cells that contain the recombinant clone must be identified. A probe is used to identify the recombinant clones.
   H. Successful genetic engineering often requires gene expression when the gene product is needed.
   I. In plant genetic engineering, genes are introduced into plant cells using the plant pathogen, *Agrobacterium*.

III. Southern Blot
   A. The Southern blot technique is a multistep procedure to identify specific sequences of DNA and to determine the size of the restriction fragments.
   B. The DNA must first be digested and then separated according to size using agarose gel electrophoresis.
   C. The DNA is denatured and transferred to a durable membrane support.
   D. Complementary sequences are identified by hybridization with a probe.

IV. Polymerase Chain Reaction
    A. The polymerase chain reaction allows efficient duplication or amplification of specific pieces of DNA.
    B. Repeated temperature cycles amplify the target DNA.
    C. Ingredients in the PCR include
        1. The target DNA or the DNA to be amplified
        2. A heat-stable DNA polymerase
        3. Each of the four DNA nucleotides
        4. Small DNA segments called primers

VOCABULARY:TERMS AND DEFINITIONS

The following list contains new terms introduced in this chapter. Use these terms to fill-in the blanks of the sentences that follow and you will have a definition or description of each new term.

DNA typing   **genetic engineering**   **ethidium bromide**
Southern blot   **probe**   **recombinant DNA**
restriction   **agarose**   **interferon**
gene cloning   **gene library**

1. The gel used to separate DNA molecules is called _____ .

2. _____ is the analysis of DNA that is analogous to blood typing which is becoming accepted for the identification of specific individuals in murder and rape cases.

3. _____ can be defined as DNA molecules which contain pieces of DNA not normally found together.

4. The production of cells with altered properties due to genes that have been inserted from other sources is known as _____ .

5. Substances found in human blood that have a variety of activities against viruses and which can be produced in genetically engineered cells is called _____ .

6. A fluorescent dye called _____ that binds to nucleic acid molecules and is used to detect these molecules in gels.

7. _____ is a technique in which genes are removed from a cell and propagated in a new host cell where it multiplies.

8. The technique in which a probe is allowed to hybridize with a single-stranded DNA molecule to identify its complement is known as _____ .

9. _____ enzymes recognize and attach to specific base sequences in double-stranded DNA and break the bonds at these points.

10. A _____ is a molecule of either DNA or RNA which has the same nucleotide sequence as a gene you are trying to identify.

11. The entire set of cloned fragments from the genome of an organism is called a _____ .

SELF-TEST OF READING MATERIAL

1. Which of the following organisms is used to introduce new genes into plants?
   a. *E. coli*
   b. *Agrobacterium*
   c. His·
   d. *Salmonella*
   e. *Bacillus*

2. The genes for insulin synthesis which are cloned into *Escherichia coli* came from
   a. pigs
   b. *Salmonella*
   c. calf pancreas
   d. *Escherichia coli*
   e. humans

3. Which of the following processes does not result in recombination?
   a. transformation
   b. conjugation
   c. transduction
   d. replication
   e. Two of the above do not.

4. Which of the following can now be produced in microorganisms?
   1. insulin
   2. human growth hormone
   3. deoxyribonuclease
   4. blood clotting factors
   5. interferon

   a. 1,2,4,5
   b. 1,2,4
   c. 1,2,5
   d. 1,2
   e. 1,2,3,4,5

5. One of the most widely used natural insecticides are the toxins produced by
   a. *Bacillus thuringiensis.*
   b. *Salmonella.*
   c. *Bacillus stereothermophilis.*
   d. *Bacillus cereus.*
   e. *Escherichia coli.*

6. Which of the following techniques allows an investigator to start with a single piece of DNA and produce billions of copies in a matter of hours?
   a. transduction
   b. gene cloning
   c. polymerase chain reaction
   d. recombinant technology
   e. restriction analysis

7. Plasmids and viruses are the most common vectors used for gene cloning.
   a. True
   b. False

8. "Scissors" that are used to cut DNA into specific pieces at defined sites are the
   a. repair enzymes.
   b. SOS.
   c. unrestricted endonucleases.
   d. restriction endonucleases.
   e. exonucleases.

9. A probe used to identify a gene can be
   a. DNA
   b. a protein.
   c. RNA
   d. a and b.
   e. a and c.

10. Machines are now available that can synthesize nucleic acids of any sequence that an investigator might desire.
    a. True
    b. False

REVIEW QUESTIONS

1. Describe four advances in the last twenty years that have been instrumental in the development of biotechnology.

2. Outline the steps involved in gene cloning.

3. Describe a gene library.

4. What are two advantages of using natural insecticides such as the *B. t.* toxins instead of chemical insecticides.

5. Why is insulin produced by gene cloning better than that isolated from animal sources?

ANSWERS:

    Vocabulary:Terms and Definitions

        1. agarose 2. DNA typing 3. recombinant DNA 4. genetic engineering 5. interferon 6. ethidium bromide
        7. gene cloning 8. Southern blot 9. restriction 10. probe 11. gene library

    Self-Test of Reading Material

        1. b 2. e 3. d 4. e 5. a 6. c 7. a 8. d 9. e 10. a

NOTES

9 CONTROL OF MICROBIAL GROWTH

The control of microbial growth is essential not only for health reasons but also for the preservation of the foods we eat, the clothes we wear, even the preservation of the homes where we live. Some methods of control, such as the preservation of food by drying or salting, are as old as civilization. Most methods of control are directed towards the most difficult forms of bacteria to kill, the endospores. A number of chemical or physical agents are available, but heat is probably the most common application. It is essential to understand the difference between an agent that is bacteriostatic and one that is bactericidal. Failure to understand this could have very serious consequences.

KEY CONCEPTS

1. Sterilization is the process of removing or killing all microorganisms and viruses in or on a material. Disinfection, by contrast, implies only a reduction in the number of infectious agents to a point where they no longer present a hazard.
2. Time, temperature, growth stage of the organism, nature of the suspending medium and the numbers of organisms present must all be considered when determining which sterilization or disinfection technique to employ.
3. Sterilization and disinfection can be accomplished by using heat, filtration, chemicals, or radiation.
4. Preservation techniques slow or halt the growth of organisms to delay spoilage.

SUMMARY

I. Approaches to Control
   A. Sterilization is the process of killing or removing all microorganisms and viruses in or on a material.
   B. Disinfection reduces the number of potential disease-causing bacteria or viruses until they no longer pose a hazard.

II. Principles Involved in Killing Microorganisms
   A. During sterilization only a fraction of the microorganisms or viruses die during a given time interval.
   B. The time to achieve sterility depends on the number of organisms present at the beginning.
   C. Different microorganisms and viruses vary in their susceptibility to sterilizing and disinfecting procedures.
   D. Numerous conditions alter the effectiveness of agents that kill microorganisms and viruses.

III. Using Heat to Kill Microorganisms and Viruses
   A. Dry heat requires more time than wet heat to kill microorganisms.
   B. Boiling kills vegetative cells and viruses but is not reliable for killing endospores.
   C. Pressure cookers and autoclaves are effective sterilizers if properly used.
   D. Pasteurization involves controlled heating at temperatures below boiling. It is not a sterilizing procedure, but eliminates many potential pathogens and delays spoilage.

IV. Sterilization by Filtration
   A. The effectiveness of filters depends on multiple factors such as pore size, chemical nature of the suspending fluid, and the amount of pressure to transfer fluid across the filter.
   B. Some membrane filters have such a small pore size that they are able to filter viruses from a suspension.

V. Chemicals Used for Sterilization, Disinfection, and Preservation
   A. The germicidal activity of chemicals can be ranked.
   B. Germicidal chemicals used in medical fields are federally regulated.

    C. A number of chemical families provide useful germicides.
       1. Alcohols rapidly kill vegetative bacteria and fungi, but are not effective against endospores.
       2. Chlorine and iodine are widely used disinfecting agents.
       3. Aldehydes include formaldehyde and glutaraldehyde.
       4. Phenolics include a large group of compounds chemically related to phenol (carbolic acid).
       5. Quaternary ammonium compounds are surface-active agents.
       6. Metals such as mercury and silver compounds were, for many years, widely used as disinfectants and antiseptics.
       7. Ethylene oxide is an extremely useful gaseous sterilizing agent with many commercial and medical applications.

VI. Radiation
    A. Gamma rays cause biological damage by producing hyperreactive ions. Ionizing radiation provides an alternative sterilizing method to ethylene oxide and heat, and has the advantages of not significantly altering the sterilized material and allowing immediate use of the object without airing.
    B. Ultraviolet radiation damages nucleic acids. It's use is limited to air and clean surfaces.
    C. Microwaves do not kill microorganisms directly.

VII. Preservation Inhibits the Growth of Microorganisms
    A. Chemicals are useful as preservatives.
    B. Growth of spoilage microorganisms is slowed or halted by low-temperature treatment.
    C. Desiccation (drying), or adding sugar and salt prevents microbial growth by reducing the water activity.

VOCABULARY:TERMS AND DEFINITIONS

The following list contains new terms introduced in this chapter. Use these terms to fill-in the blanks of the sentences that follow and you will have a definition or description of each new term.

        ionizing   **sterilization**   **antiseptic**
        sanitize   **decimal reduction**   **bacteriostatic**
        terminal cleaning   **disinfection**   **pasteurization**
        tincture

1. Gamma rays are an example of _____ radiation that causes biological damage by producing hyperreactive ions.

2. _____ is a term that indicates an antibacterial agent is primarily inhibits growth without substantial killing.

3. The complete killing or removal of all microorganisms and viruses from a material is called _____ .

4. Operating rooms and other patient care rooms are thoroughly cleaned and disinfected after use, a process known

    as _____ .

5. An _____ is a disinfectant that is compatible with human tissues.

6. According to public health standards, to _____ is to not only to substantially reduce the numbers of microorganisms on an object, but also to produce a product that has a clean appearance.

7. _____ is the process that reduces the number of potential pathogens on an object until they no longer represent a hazard.

8. An alcoholic solution of a substance is called a _____ .

9. The time required for 90% of the organisms in a population to be killed at a given temperature is called

   the ——————————————————— time.

10. The process of controlled heating of substances at temperatures below boiling which is directed toward the elimination

    specific pathogens is called ——————————————— .

SELF-TEST OF READING MATERIAL

1. Which of the following would be the fastest way to kill endospores?
   - a. autoclave at 121°C
   - b. tincture of soap
   - c. hot air oven at 180°C
   - d. chlorox solution
   - e. None of the above methods will kill endospores.

2. A substance used to disinfect human tissues is called a(n)
   - a. antibiotic.
   - b. disinfectant.
   - c. antiseptic.
   - d. sterilizer.
   - e. germicide.

3. Which of the following causes the formation of thymine dimers in cellular DNA?
   - a. X-rays
   - b. base analogs
   - c. gamma rays
   - d. nitrous acid
   - e. ultraviolet light

4. Which concentration of ethyl alcohol would is the most effective disinfectant?
   - a. 100%
   - b. 70%
   - c. 60%
   - d. 50%
   - e. 30%

5. Which of the following best describes how cells in a population die when exposed to an antimicrobial agent?
   - a. It depends on the species.
   - b. The cells all die at once.
   - c. It depends on the antimicrobial agent.
   - d. The cells are never all killed.
   - e. The cells in a population die exponentially.

6. Warmer temperatures generally interfere with the effectiveness of a disinfectant.
   - a. True
   - b. False

7. A chlorox solution and ethyl alcohol are examples of
   - a. antiseptics.
   - b. sterilizing agents.
   - c. disinfectants.
   - d. antibiotics.
   - e. base analogs.

8. If a population of bacteria is exposed to steam at 121°C, the entire population will die at once because of the moist heat at this temperature.
   - a. True
   - b. False

9. A disinfectant is an antimicrobial agent which can be used on inanimate objects while an antiseptic is an antimicrobial agent which can be used on living tissue.
   - a. True
   - b. False

10. Tincture is a solution of
    - a. iodine.
    - b. chlorox.
    - c. soap.
    - d. alcohol.
    - e. water.

11. Bacteriostatic substances will kill bacteria if allowed to remain in contact with them for long enough period of time.
    - a. True
    - b. False

12. Which of following is the least effective disinfectant?
    - a. soap
    - b. alcohol
    - c. iodine
    - d. cresol
    - e. quats

13. Which of the following can be used for sterilizing materials?
    - a. alcohol
    - b. phenolic compounds
    - c. ethylene oxide
    - d. chlorine
    - e. soap

14. Which of the following is most effective in killing bacteria?
    - a. boiling at 100°C
    - b. an autoclave at 121°C
    - c. freezing and thawing several times
    - d. freezing at -20°C
    - e. a hot air oven at 180°C

REVIEW QUESTIONS

1. Distinguish between decontamination and disinfection.

2. Why can scrubbing prior to disinfection be helpful?

3. Which microorganisms are most resistant to chemical disinfectants and which are the least resistant?

4. During sterilization, if half of the bacteria present in a suspension are killed in the first minute, what fraction of those remaining are killed during the second minute? Does it matter what method of sterilization is being used?

5. If, in New Orleans, it takes 5 minutes to sterilize a suspension of bacterial cells by boiling, would you predict it would take more or less time to accomplish the same sterilization in Denver? Explain your answer.

ANSWERS:

Vocabulary:Terms and Definitions

1. ionizing 2. bacteriostatic 3. sterilization 4. terminal cleaning 5. antiseptic 6. sanitize 7. disinfection 8. tincture 9. decimal reduction 10. pasteurization

Self-Test of Reading Material

1. a 2. c 3. e 4. b 5. e 6. b 7. c 8. b 9. a 10. d 11. b 12. a 13. c 14. b

NOTES

10 CLASSIFICATION AND IDENTIFICATION OF BACTERIA

A system for the classification of microorganisms is essential so that microbiologists can keep things straight. A system of classification helps microbiologists organize the volumes of information that exist on a great many microbes into a manageable, logical scheme. Ideally, such a classification system would also show evolutionary relationships among the microorganisms. That is, it would illustrate how one organism is related to another. The methods of biotechnology now hold the promise of reaching the goal of a natural system of classification. A patient, sick from an infectious disease, however, does not care particularly what species of bacteria they have - they are just interested in getting well. Their recovery, however, often depends on the proper identification of the infectious agent.

KEY CONCEPTS

1. The best classification schemes group organisms that are related through evolution and separate those that are unrelated.
2. At the present time, the classification scheme for bacteria is based not on evolution but on convenience, and consists of readily determined characteristics such as shape and type of metabolism.
3. The most convenient method of determining the evolutionary relatedness between organisms is based on comparing the sequence of nuceoltides in their DNA.
4. Unexpected relationships between various bacteria have been revealed by comparing their 16S ribosomal RNA sequences.
5. In clinical laboratories, identifying the genus and species of an organism is more important than understanding its evolutionary relationship to other organisms.

SUMMARY

I. Problems in Classifying Organisms
    A. The primary purpose of any classification scheme is to provide easy identification of an organism.
    B. The best classification schemes are based on evolutionary relatedness.
    C. Species is the basic taxonomic unit in bacteria; two species differ from one another in several features determined by genes.

II. Present Day Classification Schemes
    A. Descriptions of bacteria are contained in a reference text, *Bergey's Manual of Systematic Bacteriology*, which is published in four volumes.

III. Numerical Taxonomy
    A. This technique measures the relatedness between organisms by determining how many characteristics the organisms have in common.

IV. Molecular Approaches to Taxonomy
    A. A comparison of the DNA base composition of organisms indicates if they are related.
        1. If the GC content of two organisms differs by more than a few percent, they cannot be related.
        2. Having the same GC content does not necessarily mean that two organisms are related.
    B. A comparison of the sequence of bases in DNA can help determine the relatedness of organisms.
        1. Nucleic acid hybridization is the most widely used and accurate way to determine how organisms are related to one another.
    C. Similarities in amino acid sequences in the same protein can indicate relatedness between organisms.
        1. Only a minute portion of the total genetic information of the cell is measured by this method.

    D. A comparison of the base sequence in 16S ribosomal RNA reveals evolutionary relationships between diverse groups of bacteria.

        1. This technique allows evolutionary relations to be seen between organisms that are not demonstrated by nucleic acid hybridization.

        2. Certain sequences in 16S ribosomal RNA are always found in one of three groups of organisms and identify the organism as belonging to the Eubacteria, Archaea or Eukarya.

V.   Methods of Identifying Bacteria

    A. The more tests that are done, the more accurate the identification.

    B. There are four major approaches: direct techniques, culture techniques, detection of microbial by-products, molecular biological techniques.

    C. Direct methods of identification involve microscopic examination of specimens without culturing the organisms.

        1. Wet mounts are made.

        2. Stains, especially the Gram stain, are useful if enough organisms are present.

        3. Many harmful bacteria cannot be distinguished from nonharmful bacteria by their appearance.

    D. Culture techniques involve growing the organisms in pure culture.

        1. The specimen that is cultured depends on symptoms of disease.

        2. Culturing specimens may be necessary if infectious agents are present in small numbers.

        3. Isolated organisms can be precisely identified.

    E. The detection of the products of metabolism is sometimes easier and faster than culturing the organism.

        1. By-products are measured by assaying for breakdown products of labeled energy sources.

        2. This technique allows faster identification of the organisms and therefore faster treatment.

    F. Molecular approaches in diagnostic microbiology usually involve analysis of DNA.

        1. Nucleic acid hybridization between the DNAs of two organisms, one known, the other unknown, will indicate whether or not they are related.

        2. Comparison of the cleavage pattern of two DNA samples, one from a known, another from an unknown organism, by the same restriction enzyme will indicate whether or not they are related.

        3. DNA probes can be used to identify species of bacteria in their natural environment or after cultivation.

VI. Identifying Organisms That Cannot Be Cultivated

    A. Microorganisms that cannot be cultured can still be identified.

        1. One technique is to copy a piece of DNA by PCR that is specific to one of the major cell domains (Eubacteria, Archaea or Eukarya) and compare the sequence with the signature sequences from previously identified organisms.

        2. PCR technology can be used to detect the presence of a single organism or virus particle in any environment.

VOCABULARY: TERMS AND DEFINITIONS

The following list contains new terms introduced in this chapter. Use these terms to fill-in the blanks of the sentences that follow and you will have a definition or description of each new term.

strain	**diagnostic**	**GC content**
numerical	**classification**	**species**
taxonomy	**nucleic acid hybridization**	

1. The technique for comparing the sequences of bases in DNA to determine the relatedness of organisms is

    termed _____ .

2. _____ is the arranging of organisms into related groups.

3. If two organisms differ only in minor phenotypic properties, then one organism may be considered a

_____ rather than another species.

4. _____ is the science of classification.

5. The DNA base composition of an organism is usually expressed as the percent of guanine plus cytosine and is

known as the _____ .

6. _____ microbiology is the area of microbiology concerned with the identification of infectious
agents.

7. The _____ is the basic taxonomic unit for all organisms, but its definition is different for
microorganisms compared to plants and animals.

8. An approach to taxonomy in which a large number of characteristics are compared and a percentage of common

characteristics is calculated to show relatedness is called _____ taxonomy.

SELF-TEST OF READING MATERIAL

1. The GC content of two organisms was found to be 23% and 24%. Which of the following statements is true?
   a. The two organisms are closely related to each other.
   b. The two organisms are not closely related to each other.
   c. The two organisms are closely related only if they belong to the same taxonomic group.
   d. The similarity in GC contents may be coincidental and the organisms are not related.
   e. Two of the above are correct answers.

2. Bacteria are not given species names because they do not reproduce sexually and do not meet the definition of a
   species.
   a. True
   b. False

3. The ideal system of classification would be based on
   a. structure.
   b. easily observed characteristics.
   c. phenotypes.
   d. evolutionary relationships.
   e. PCR.

4. A surprising result of the analysis of 16S ribosomal RNA demonstrated that
   a. archaea are a separate group from the prokaryotes.
   b. prokaryotes and eukaryotes actually belong to the same group.
   c. bacteria are closely related to certain plants.
   d. archaea are a subgroup within and closely related to the prokaryotes.
   e. there are as many as a dozen different groups very distinct from each other.

5. Analysis of the sequence of amino acids in proteins gives _____ analysis of nucleotide sequences in DNA.
   a. considerably more information about the genetic composition of an organism than
   b. approximately the same information about the genetic composition of an organism as
   c. considerably less information about the genetic composition of an organism than

6. Direct examination for the identification of an infectious agent is more useful for many parasitic diseases than it is for
   bacterial diseases.
   a. True
   b. False

7. When identifying an unknown organism
    a. a few selected tests are usually preferred for an accurate identification.
    b. culturing the organism is necessary before an identification can be made.
    c. the more tests that can be made the better.
    d. only the techniques of molecular biology are accepted as valid.
    e. Two of the above answers are correct.

8. The Gram stain is not sensitive or specific enough to be of any practical use in the clinical laboratory.
    a. True
    b. False

REVIEW QUESTIONS

1. Based on a comparison of the 16S ribosomal RNA, what are the three distinct cell types?

2. Distinguish between an artificial system of classification and a natural system.

3. How is a species defined for bacteria compared to animals?

4. Why are DNA probes such powerful tools in classification and identification of bacteria?

ANSWERS:
   Vocabulary:Terms and Definitions
      1. nucleic acid hybridization 2. classification 3. strain 4. taxonomy 5. GC content 6. diagnostic 7. species
      8. numerical
   Self-Test of Reading Material
      1. e 2. b 3. d 4. a 5. c 6. a 7. c 8. b

11 *PROKARYOTIC MICROORGANISMS: EUBACTERIA AND ARCHAEA*

This chapter is a survey of the major groups of prokaryotic microorganisms as outlined in the Bergey's Manual of Systematic Bacteriology. In an introductory course, students of microbiology come to think of bacteria only in terms of the garden variety types such as *E. coli*. This would be like thinking of birds as all being sparrows. The diversity of bacteria is not only extreme, but also, even beautiful. Bacteria that kill other bacteria; bacteria with stalks; bacteria that look like fungi; and bacteria that live where we thought nothing could live are only a few examples of those introduced in this chapter.

KEY CONCEPTS

1. Prokaryotes show extreme diversity in appearance, growth requirements, and metabolism.
2. Some prokaryotes prey on other prokaryotes.
3. Some prokaryotes can multiply only as symbionts within eukaryotic cells.
4. Prokaryotes are vitally important in the recycling of carbon, nitrogen, and sulfur.
5. Although they have a prokaryotic cell structure, archaea are probably as closely related to eukaryotes as to eubacteria.

SUMMARY

I. Prokaryotes With a Gram-negative Type of Cell Wall
   A. The spirochetes
      1. Spirochetes are flexible and have a spiral shape. Axial filaments contribute to their motility.
   B. Aerobic, motile, helical or comma-shaped gram-negative bacteria.
      1. This group of spiral and comma-shaped gram-negative bacteria have a rigid cell wall, move by means of flagella, and require oxygen for growth.
      2. Microaerophiles require an atmosphere containing oxygen, but a much lower concentration than air.
      3. *Bdellovibrios* prey on other gram-negative bacteria by reproducing between the cell wall and cytoplasmic membrane.
   C. Gram-negative aerobic rods and cocci
      1. This large, diverse group of gram-negative bacteria is characterized by rod or coccoid shape and aerobic growth.
      2. *Rhizobium* species are important in converting atmospheric nitrogen (N_2) to a form usable by plants (nitrogen fixation). They live within the root cells of many plant species.
      3. Pseudomonads often produce non-photosynthetic pigments can color the medium green in some cases. They utilize many complex organic compounds. Some can use nitrate (NO_3^-) as a substitute for molecular oxygen (O_2), an example of anaerobic respiration.
      4. Azotobacteria are soil organisms that can fix nitrogen under aerobic conditions.
   D. Facultatively anaerobic Gram-negative rods
      1. Facultatively anaerobic gram-negative rods grow best aerobically but can also grow anaerobically if a suitable organic compound is present.
      2. Enterobacteria include *Escherichia coli,* universally present in the human intestine.
      3. Vibrios and related bacteria are mostly marine organisms. Some possess luciferase and are luminescent.
   E. Anaerobic Gram-negative straight, curved and helical rods
      1. Anaerobic, gram-negative fermentative rods have various shapes and are inhibited or killed by oxygen. Some species inhabit the body cavities of humans.
   F. Dissimilatory sulfate- or sulfur-reducing bacteria
      1. Dissimilatory sulfate- or sulfur-reducing bacteria utilize sulfur or oxidized sulfur compounds as a final electron acceptor. They are important in recycling sulfur .

G. The Rickettsias and chlamydias
    1. Rickettsias and chlamydias are generally unable to grow outside the cells of a host animal but members of the genus *Bartonella* are an exception. Rickettsias are transmitted from one host to another by insects, mites, and ticks.
    2. Chlamydias multiply in a vegetative form and then differentiate into an infectious form. They are a unique group, lacking peptidoglycan in their cell walls.
H. Endosymbionts
    1. Endosymbionts live within the cells of eukaryotic hosts in a stable association that can be beneficial to both organisms. Mitochondria and some chloroplasts of eukaryotes probably evolved from prokaryotic endosymbionts.
    2. In endosymbionts of protozoa, environmental factors can upset the endosymbiotic relationship, leading, in some cases, to the death of both host and endosymbiont.
    3. Many types of endosymbionts of other eukaryotes are known, including endosymbionts of fungi, insects and parasitic worms.
I. Gliding, non-photosynthetic bacteria
    1. Gliding bacteria move slowly by an unknown mechanism when in contact with a surface. They can break down complex polysaccharides such as cellulose. One species causes a disease of fish.
J. Myxobacteria
    1. The myxobacteria are gliding, fruiting bacteria. They can produce fruiting bodies, macroscopic aggregations of resting cells called microcysts, which are more resistant to heat, drying and radiation than the vegetative cells. Chemical signals are involved in initiating and coordinating the action of the individual cells.
K. Sheathed Bacteria
    1. Sheathed bacteria are unbranching, filamentous forms in which the filament is enclosed in a sheath of lipoprotein-polysaccharide material. The sheath protects the bacteria and attaches them to solid surfaces. Sheathed bacteria may form brown scum in polluted streams, and they can plug pipes used in sewage treatment systems.
L. Anoxygenic phototrophic bacteria
    1. Anoxygenic phototrophs obtain their energy from sunlight, but do not produce oxygen as a by-product of photosynthesis. They require anaerobic growth conditions and a hydrogen donor other than water. Their ancestors probably predominated in the early history of the earth.
M. Oxygenic photosynthetic bacteria
    1. Cyanobacteria are phototrophs that, like algae and plants, liberate oxygen during photosynthesis.
    2. Their chlorophyll type and photosynthetic process are essentially the same as the algae and plants, but unlike these eukaryotes, many species of cyanobacteria can fix nitrogen.
    3. Accessory pigments called phycobiliproteins help trap light energy.
N. Budding and/or appendaged bacteria.
    1. Prosthecate and budding bacteria have projections of their cytoplasm into unusual structures. The structures function in attachment and increase the cell's surface area for absorbing nutrients
O. Aerobic chemolithotrophic bacteria.
    1. The chemolithotrophic bacteria can grow in the dark using an inorganic substance as an energy source and CO_2 for a carbon source. There are four types of chemolithotrophic bacteria: nitrifiers, sulfur oxidizers, hydrogen oxidizers, and metal oxidizers.
II. Prokaryotes With a Gram-Positive Type of Cell Wall
    A. Gram-positive cocci
        1. Gram-positive cocci may occur in a chainlike arrangement or in clusters. Aerobic, facultative and anaerobic forms exist. One group is remarkably radiation-resistant.
    B. Regular non-spore-forming gram-positive rods
        1. The non-spore-forming gram-positive rods, compose a group of mostly unicellular organisms with smooth, straight outlines.
        2. Within the genus *Lactobacillus*, some species help protect against infectious disease while others have essential roles in making pickles, yogurt and other foods.
    C. Irregular-shaped, non-spore-forming, gram-positive rods
        1. Irregular non-spore forming gram-positive rods may show misshapen cells, uneven or metachromatic staining, or branching chains of bacteria. Snapping division is common.
        2. The corynebacteria are club-shaped and may show metachromatic staining. One species causes diphtheria.

3. Arthrobacters help degrade pesticides in soil. Their life cycle includes changes in shape, ranging from rod to coccus and back to rod.
4. Members of the genus *Actinomyces* grow as branching filaments that fragment into rods.

D. Mycobacteria

1. Mycobacteria demonstrate the acid-fast staining property.. Unusual cell wall lipids make them resistant to staining by other methods and also protect them against disinfectants.
2. Certain species are responsible for tuberculosis and leprosy.

E. Actinomycetes with aerial and specialized hyphae

1. Nocardioforms are filamentous organisms that usually form aerial mycelia and fragment into bacillary forms.
2. Other Actinomycetes develop spores called conidia. Conidia are not as resistant as endospores, except in the genus *Thermoactinomyces*. Conidia may occur in sacks as in the genera *Frankia* and *Dermatophilus*. Streptomycetes and others include antibiotic-producing as well as disease-causing forms.

III. Prokaryotes that Lack Cell Walls

A. The mycoplasmas

1. The mycoplasmas lack cell walls. They appear to have evolved from an ancestor similar to gram-positive bacteria. One group, the spiroplasmas, has species with a spiral shape and motility similar to spirochetes.

IV. Prokaryotes With Cell Walls that Lack Peptidoglycan

A. Archaea differ from eubacteria in many respects.

1. Archaea are only distantly related to other prokaryotes. They lack peptidoglycan in their cell walls.
2. They appear to have undergone less evolutionary change than eubacteria.
3. They exist mainly today under conditions considered adverse for eubacteria.
4. There are three kinds of archaea: methanogens, halobacteria, and thermoacidophiles.

VOCABULARY:TERMS AND DEFINITIONS

The following list contains new terms introduced in this chapter. Use these terms to fill-in the blanks of the sentences that follow and you will have a definition or description of each new term.

axial filaments	**hyphae**	**endosymbiosis**
coenocytic	**anaerobic**	**halobacteria**
ß-hemolysis	**methanogens**	**thermoacidophiles**

1. Many species of eukaryotic organisms contain gram-negative bacteria living inside their cytoplasm in a relationship

   known as ⸻ .

2. In ⸻ respiration, nitrate may be substituted for oxygen as the final electron acceptor.

3. ⸻ are flagella-like filaments that originate at the ends of a cell and overlap in the middle and are used for motility by spirochetes.

4. Three distinct groups of archaea are the ⸻ , the ⸻ ,

   and the ⸻ .

5. ⸻ are branching filaments found in actinomycetes and fungi.

6. On blood agar plates when red blood cells are completely destroyed, leaving a clear area around the colony, the effect is

   called ⸻ .

7. In some organisms, the hyphal filaments are ⸻ meaning that the cytoplasm of each cell is continuous with that of adjacent cells because they do not have cell walls separating them.

SELF-TEST OF READING MATERIAL

Match the organisms on the left with their description on the right. A description may be used more than once and an organism may have more than one description.

_____ 1.	*Neisseria gonorrhoeae*	a. spirochete
_____ 2.	Archaea	b. comma-shaped
_____ 3.	*Mycoplasma*	c. bactivorus
_____ 4.	Chlamydias	d. nitrogen-fixer
_____ 5.	*Treponema pallidum*	e. sexually transmitted disease
_____ 6.	*Bdellovibrio*	f. coccus
_____ 7.	*Campylobacter jejuni*	g. obligate intracellular organism
_____ 8.	*Vibrio cholerae*	h. salmon poisoning
_____ 9.	*Rhizobium*	i. tuberculosis
_____ 10.	*Neorickettsia*	j. walking pneumonia
_____ 11.	*Mycobacterium*	k. growth in extreme environments
_____ 12.	*Borrelia burgdorferi*	l. diarrhea and dysentery

REVIEW QUESTIONS

1. How do *Treponema pallidum, Borrelia burgdorferi* and *Leptospira interrogans* differ in their transmission to human hosts?

2. What important function do some members of the genus *Rhizobium* perform for plants?

3. In what ways do rickettsias differ from chlamydias?

4. Explain the statement, " The genus *Pseudomonas* contains species with remarkably diverse biochemical capabilities."

5. How is the hypothesis for the endosymbiotic origin of mitochondria supported by the presence of endosymbiotes in *Paramecium*?

6. Give three examples of energy sources used by chemolithotrophs.

7. Distinguish among the three groups of archaea.

8. What distinguishes mycoplasmas and spiroplasmas from other bacteria?

NOTES

ANSWERS:

Vocabulary:Terms and Definitions

1. endosymbiosis 2. anaerobic 3. axial filaments 4. methanogens/halobacteria/thermoacidophiles 5. hyphae
6. ß-hemolysis 7. coenocytic

Self-Test of Reading Material

1. e,f 2. k 3. j 4. e,g 5. a,e 6. b,c 7. b,l 8. b,l 9. d 10. g,h 11. i 12. a

12 EUKARYOTIC MICROORGANISMS: ALGAE, FUNGI, PROTOZOA

The microbial world includes eukaryotic microorganisms as well as prokaryotic microorganisms and viruses. This chapter describes the three groups of eukaryotic microorganisms, the algae, fungi, and protozoa with special emphasis on those that cause disease. The algae are not usually associated with disease except for certain types of poisoning due to toxins they produce. Members of the fungi can cause serious disease in plants and animals, including humans. Members of the protozoa are responsible for diseases such as malaria which affects the lives of millions of people.

KEY CONCEPTS

1. Algae, fungi, and protozoa are eukaryotic organisms and as such have eukaryotic cell structures. Many of these organisms are microscopic.
2. Algae are very important in the food chain and carbon cycle but are of minor importance in causing disease in humans.
3. Fungi play an important role in the decomposition of organic materials. They also cause a great deal of damage to food crops as well as other materials, and they are responsible for some serious diseases in humans.
4. Protozoa are single-celled often motile organisms that are an important part of the zooplankton and the food chain. Some species cause diseases that affect a large part of the world's population.

SUMMARY

I. Algae
   A. Algae are photosynthetic organisms that lack the complex tissue development of higher organisms.
   B. Algae are classified by photosynthetic pigments.
   C. Algae are found in a wide variety of habitats.
   D. Algae are typical eukaryotic organisms.
      1. Algal cell walls are rigid and contain cellulose.
      2. Microscopic algae are usually single cells but can also live as colonies.
      3. Macroscopic algae are multicellular with a variety of structures such as holdfasts, blades, bladders, and stipes.
      4. Reproduction in algae can be asexual or sexual. In some algae the generations alternate.
   E. Some dinoflagellates produce toxins that cause paralytic shellfish poisoning.

II. Fungi
   A. Fungi are nonmotile, non-photosynthetic, heterotrophic organisms.
   B. Fungi are primarily terrestrial and occur in a large variety of habitats.
   C. The terms yeast, mold, and mushroom refer to morphological forms of fungi.
      1. Yeasts are single cells.
      2. Molds are filamentous.
      3. Mushrooms are the large fruiting bodies of certain fungi that are often edible.
      4. Dimorphic fungi exist as either yeast or mycelium, depending on conditions.
   D. Classification of the fungi is frequently revised. We will consider them in three groups.
      1. Amastigomycota - terrestrial fungi
      2. Mastigomycota - water molds, flagellated lower fungi
      3. Gymnomycota - the slime molds
   E. Fungi cause disease in humans in four ways.
      1. Some people experience an allergic reaction to fungi.
      2. Ergot and aflatoxin are two fungal poisons that seriously affect humans.
      3. Fungal infections are known as mycoses.
      4. Fungi destroy human food supplies.

F. Fungi form symbiotic relationships with a variety of other organisms. Lichens and mycorrhizas are examples.
G. Fungi are economically important organisms..
   1. They aid in the production of foods such as cheese, bread, beer, and wine.
   2. They are a source of antimicrobial medicines.
   3. They spoil foods and degrade organic materials.
   4. They cause many plant diseases.
   5. They are used for genetic and biochemical studies.
   6. Genetically engineered yeasts produce many useful compounds such as insulin and growth hormone.

III. Protozoa
   A. Protozoa are microscopic, unicellular organisms that usually lack photosynthetic capabilities.
   B. Protozoan habitats are diverse but they all require moisture in some form.
   C. One way to group protozoa is on their modes of locomotion.
      1. Mastigophora-flagella
      2. Sarcodina-pseudopodia
      3. Sporozoa-flagella or nonmotile
      4. Ciliata-cilia
   D. Protozoa lack a cell wall but maintain their distinctive shape by other means.
   E. Protozoa display polymorphism by taking on different forms in different environments.
   F. Many protozoa take in nutrients by pinocytosis and/or phagocytosis
   G. Reproduction is by both asexual and sexual methods.
   H. Protozoa are an important part of the food chain.
   I. Protozoa cause a variety of serious diseases in humans.

VOCABULARY:TERMS AND DEFINITIONS

_____1. Another name for fungal diseases

_____2. Free-floating, photosynthetic organisms in aquatic environments

_____3. Engulfment of large food particles

_____4. Vegetative or feeding form of a protozoan

_____5. The product of the fusion of two gametes

_____6. Reproduction by multiple fissions

_____7. Poison produced by fungi that may have been responsible for the Salem witch hunts

_____8. A mass of hyphae

_____9. A structure that anchors certain algae to a solid surface

_____10. A form of asexual reproduction in filamentous algae in which a portion breaks off to form a new organism

_____11. The uptake of soluble substances through a cell membrane

a. phytoplankton
b. schizogony
c. holdfast
d. phagocytosis
e. fragmentation
f. pinocytosis
g. zygote
h. cyst
i. hypha
j. trophozoite
k. mycelium
l. lichens
m. rhizoids
n. ergot
o. mycoses

_____12.   A single filament of a mold

_____13.   The result of the association of an alga and a fungus

_____14.   Specialized hyphae that anchor fungi to a substrate

_____15.   A resting form of a protozoan

SELF-TEST OF READING MATERIAL

1. Which of the following is the most likely way to get an infection of *Giardia lamblia*?
    a. mosquito bite
    b. injection
    c. ingestion of a cyst
    d. inhalation of an endospore
    e. ingestion of a trophozoite

2. Malaria is caused by which of the following organisms?
    a. *Toxoplasma*
    b. *Trypanosoma*
    c. *Trichomonas*
    d. *Giardia*
    e. *Plasmodium*

3. Malaria is transmitted to humans by
    a. ingesting cysts in water or food.
    b. inhaling trophozoites.
    c. mosquito bites.
    d. dirty fingers.
    e. handling cats.

4. A unique feature of the Sporozoans is that they
    a. divide by transverse binary fission.
    b. reproduce exclusively by sexual means.
    c. are not motile in the adult form.
    d. reproduce exclusively by asexual means.
    e. have two types of nuclei.

5. Which of the following genera causes vaginitis?
    a. *Toxoplasma*
    b. *Giardia*
    c. *Plasmodium*
    d. *Trichomonas*
    e. *Trypanosoma*

6. Which of the following statements describe protozoans?
    a. They are eukaryotes.
    b. They can be grouped on the basis of their means of motility.
    c. They reproduce only by asexual means.
    d. All of the above are correct statements.
    e. Only a and b are correct.

7. Which of the following statements about the Deuteromycetes are true?
    a. Their usual habitat is aquatic.
    b. Sexual reproduction is absent or unknown.
    c. Their cell walls are cellulose.
    d. All of the above statements are correct.
    e. a and b are correct statements.

8. Which of the following statements describe fungi?
    a. They are eukaryotes.
    b. Most fungi are anaerobic.
    c. They reproduce only by sexual means.
    d. All of the above statements are correct.
    e. a and b are correct statements.

9. Fungal infections of the skin are called
    a. systemic mycoses.
    b. superficial psychoses.
    c. intermediate mycoses.

    d. superficial mycoses.
    e. yeasts.

10. Which of the following statements describe algae?
    a. They are eukaryotes.
    b. They may be either microscopic or macroscopic.
    c. They are classified according to their photosynthetic pigments.

    d. All of the above statements are correct.
    e. a and b are correct statements.

REVIEW QUESTIONS

1. Compare the algae, fungi, and protozoa . What characteristics do all these organisms have in common? What features of each cause them to be separated into distinct classification groups?

2. What characteristics of algae account for their general lack of disease causing properties?

3. What properties of fungi account for their ability to be so destructive?

4. Although we think most often of protozoa as disease causing organisms what would the world be like if there were no protozoa?

NOTES

ANSWERS:
  Terms and Definitions
    1. o 2. a 3. d 4. j 5. g 6. b 7. n 8. k 9. c 10. e 11. f 12. i 13. l 14. m 15. h
  Self-Test of Reading Material
    1. c 2. e 3. c 4. c 5. d 6. e 7. b 8. a 9. d 10. d

13 THE NATURE AND STUDY OF VIRUSES

Viruses are very different from any of the cellular organisms that we have described for you. Their organization is at the macromolecular level, several steps below the organization of a cell, eukaryotic or prokaryotic. Viruses are little more than a few genes carried in a protein coat. They have no metabolic machinery of their own, and therefore, must rely upon that of their host cell. Viruses cannot replicate or reproduce outside of a host cell. Viruses infect both eukaryotic and prokaryotic cells. Much of our knowledge of viral replication is from studies of viral infections of bacteria. Viruses can confer new properties on their host cells.

KEY CONCEPTS

1. Viruses are a set of genes enclosed in a protein coat.
2. In order to multiply, viruses depend, to varying degrees, on the enzymes of the cells they invade.
3. Viruses that infect bacteria serve as excellent models for viruses infecting animal and plant cells.
4. Viruses can replicate within cells that they subsequently lyse or they can become integrated into host cell DNA and confer new properties on the cells.
5. Different viruses can infect and replicate within all kinds of cells, both prokaryotic and eukaryotic.

SUMMARY

I. Viruses - General Characteristics
   A. Viruses have several features that distinguish them from cells.
      1. They are very small - 100 to 1,000 times smaller in diameter than the cells they invade.
      2. They contain very few genes.
      3. The nucleic acids can be either DNA or RNA, but never both.
      4. Viral proteins and nucleic acids replicate separately.

II. Virus Architecture
   A. Viruses can have different shapes. Some are polyhedral (having many flat surfaces), others are helical, and others are a combination of both.
   B. The shape of the virus is determined by the shape of its protein coat (capsid).
   C. Some viruses have only a capsid whereas some animal viruses have a lipid membrane (envelope) surrounding the capsid.

III. Virus - Host Relationships as Exemplified by Bacteriophage - Bacterial Interactions.
   A. Three major types of virus - host relationships are recognized.
      1. A productive response in which more copies of virus are made and these are released following lysis of the host cell.
      2. A productive response in which the virus leaves the cell without lysing it.
      3. A latent state in which the viral nucleic acid becomes integrated into the host chromosome.

IV. Virus Replication in a Productive Infection Resulting in Cell Lysis
   A. Step 1 - Phages attach to specific sites on host bacterium.
   B. Step 2 - Viral nucleic acid enters the host cell; protein coat remains on the outside.
   C. Step 3 - Phage DNA is transcribed sequentially leading to synthesis of specific proteins.
   D. Step 4 - Phage DNA is replicated and phage structural proteins are synthesized independently.
   E. Step 5 - Phage DNA and protein assemble to form mature virions - the maturation process.
   F. Step 6 - Virions are released from the host cell by lysis of the bacterial cell wall.

V. Phage Growth: A One-Step Growth Curve
   A. This procedure involves assaying the number of virions inside and outside the bacteria at various times after adsorption.
   B. Samples are removed at various times, one set treated with chloroform, the other is not. Samples are inoculated onto a lawn of bacteria. The sites on the lawn at which phages infect bacteria will cause the bacteria to lyse and leave a clear area known as a plaque.

VI. Viral Replication in a Latent State
   A. The DNA of temperate viruses is integrated into the host chromosome.
   B. Integration of phage DNA into the bacterial chromosome occurs because of identical DNA sequences in the phage and the bacteria.
   C. Repressor must function continuously to keep prophage in the integrated state.
   D. Activation of SOS repair systems destroys the repressor; the phage DNA is excised and virions are produced that lyse the cells.
   E. Lysogens are immune to infection by the same phage.
   F. Prophage often codes for proteins which alter the properties of the host (lysogenic conversion).
   G. Virus can be released without host cell lysis.
      1. Viruses, such as filamentous phages, are released without killing the host cell. They leave the cell by extrusion.
      2. The virus is assembled as it leaves the host cell.

VII. Transduction
   A. Both virulent and temperate phages can transfer DNA from one host to another host.
      1. In generalized transduction, virulent phages can transfer any part of the genome.
      2. In specialized transduction, temperate phages can transfer only a specific set of genes.

VIII. Host Range of Viruses
   A. Any given virus can generally infect cells of only one or a few species.
   B. Limitation in host range is usually because animal and bacterial viruses must adsorb to specific receptors on the host cell surface.

IX. Methods Used to Study Viruses
   A. Many eukaryotic host cells can be cultivated in the laboratory with the techniques of tissue culture.
   B. Normal cells from a vertebrate animal grow as a monolayer attached to the surface of a petri dish, but for only a limited number of generations before they die.
   C. Viruses can infect, multiply and lyse animal cells growing in a monolayer, resulting in readily observable clearings (plaques).
   D. Infected tissue culture cells undergo characteristic changes in their appearance depending on the virus.
   E. Quantal assays measures the number of infective virions by their effect on different host cell systems.
   F. The presence of some viruses can be detected because they can cause red blood cells to clump.

VOCABULARY:TERMS AND DEFINITIONS

_____ 1. Phages that can either replicate or can integrate into the host chromosome

_____ 2. Viruses that have a lipid membrane acquired during release from the host cell

_____ 3. Viruses that infect bacteria

_____ 4. Changes in the appearence of virus infected tissue culture cells

_____ 5. Process by which any gene can be transferred by a phage from one bacterium to another

_____ 6. Identical protein subunits composing the capsid of a virus

_____ 7. An area of clearing on a bacterial lawn indicative of a phage infection

_____ 8. The technique of cultivating animal cells in the laboratory

_____ 9. A bacterial cell carrying a prophage

_____ 10. Protective coat surrounding a virus nucleic acid core

_____ 11. An enzyme located in the tail of some phages that digests a small hole in the bacterial cell wall

_____ 12. A complete and infective virus particle

_____ 13. Genetic elements that can replicate as part of the chromosome or independently of it

_____ 14. Phage DNA integrated into the host cell DNA

_____ 15. Proteinaceous infectious agents

a. bacteriophages
b. virion
c. capsid
d. capsomeres
e. enveloped viruses
f. prion
g. lysozyme
h. plaque
i. temperate phage
j. prophage
k. lysogen
l. generalized transduction
m. episomes
n. tissue culture
o. cytopathic effect

SELF-TEST OF READING MATERIAL

1. A bacteriophage after it becomes integrated into its host's chromosome is called a(n)
   a. prophage
   b. temperate phage
   c. lysogen
   d. intemperate phage
   e. lytic

2. List the following stages of the lytic cycle of bacteriophages in the proper sequence.
   1. phage DNA synthesized
   2. lysis
   3. penetration
   4. attachemnt
   5. phage DNA transcribed
   6. assembly

   a. 3,5,1,2,4,6
   b. 4,5,1,6,2
   c. 5,6,2,4,3,1
   d. 4,5,3,1,6,2
   e. 4,3,5,1,6,2

3. A complete infective virus particle is called a
    a. bacteriophage.
    b. viroid.
    c. prophage.
    d. lysogen.
    e. virion.

4. All viruses, except the bacteriophages, are obligate intracellular parasites.
    a. True
    b. False

5. In viruses that have envelopes, the origin of these envelopes is
    a. viral.
    b. host.
    c. bacterial.
    d. phage.
    e. partially viral and partially host.

6. A virion is a
    a. naked, infectious piece of RNA.
    b. complete, infectious virus particle.
    c. capsid without a nucleic acid.
    d. nucleic acid without a capsid.
    e. a naked, infectious piece of DNA.

7. All viruses require a host cell for replication.
    a. True
    b. False

8. The transfer of genetic information which is mediated by a bacteriophage is called
    a. translation.
    b. transduction.
    c. transfusion.
    d. transformation.
    e. transcription.

9. Viruses, except for bacteriophages, contain a single type of nucleic acid.
    a. True
    b. False

10. Recent breakthroughs in techniques have resulted in the cultivation of viruses in cell-free medium in the lab.
    a. True
    b. False

REVIEW QUESTIONS

1. Describe four features that distinguish viruses from cells.

2. Viruses have a limited number of shapes. Why is this?

3. What is lysogenic conversion? Give an example.

4. Compare and contrast the types of transduction.

ANSWERS:
  Vocabulary:Terms and Definitions
    1. i 2. e 3. a 4. o 5. l 6. d 7. h 8. n 9. k 10. c 11. g 12. b 13. m 14. j 15. f
  Self-Test of Reading Material
    1. a 2. e 3. e 4. b 5. e 6. b 7. a 8. b 9. b 10. b

NOTES

14 *VIRUSES OF ANIMALS AND PLANTS*

Although much of the biology of bacteriophages applies to animal and plant viruses, there are significant differences in viral infections of eukaryotic cells. This chapter explores some of those differences with an emphasis on animal viruses. You will be introduced to the classification of viruses and then a description of animal virus replication. Animal and plant viruses cause damage to their host cells in a variety of ways including tumor formation. The human immunodeficiency virus (HIV) is the causative agent of AIDS or the acquired immunodeficiency syndrome.

KEY CONCEPTS

1. Viruses are classified on the basis of their morphology, size, and the physical and chemical nature of their nucleic acid.
2. The interactions of animal viruses with their hosts are similar to the interactions between bacteriophages and bacteria.
3. The replication of viral nucleic acid depends to varying degrees on the enzymes of nucleic acid replication of the host cell.
4. Symptoms of viral-caused disease result from tissue damage of the host.
5. Some animal tumors result from viral DNA becoming integrated into the host genome and coding for the synthesis of abnormal proteins, which interfere with normal control of host cell growth.

SUMMARY

I. Classification Scheme of Animal Viruses
   A. Animal viruses are classified into families, genera and species based mainly on four characteristics
      1. Size
      2. Nature of nucleic acid they contain
      3. Presence or absence of an envelope
      4. Shape of virus particle
   B. Viruses can also be grouped according to their route of transmission.
      1. Enteric viruses contaminate food and water and typically infect and replicate in the intestinal tract.
      2. Respiratory viruses enter the body on inhaled droplets and replicate in the respiratory tract.
      3. Arboviruses are transmitted to hosts by the bites of arthropods.
      4. Sexually transmitted viruses cause lesions in the genital tract.

II. Virus Interactions with Animal Cells
   A. Three kinds of viral-host cell interactions can be identified:
      1. productive infection which results in cell lysis
      2. persistent infections in which the virus often buds from the cell without killing it
      3. latent infections in which viral nucleic acid is integrated into the host cell genome

III. Virus multiplication
   A. The first step in infection is adsorption of the virus to the host cell.
      1. Attachment proteins of the virus bind to specific host cell receptors.
      2. The host cell is resistant if it lacks receptors to the virus.
   B. Viral entry often involves fusion of the cytoplasmic membrane of the host with the virion envelope
      1. Naked virions enter by endocytosis
   C. The replication of viral nucleic acid depends to varying degrees on enzymes of the host cell
      1. Larger viruses generally encode more enzymes and are less dependent on host cell enzymes.
   D. Maturation of virions involves self assembly of viral component parts
   E. Virions are released as a result of lysis of dead cells or by budding from living cells.
      1. Animal and plant viruses do not use lysozyme

IV. Interactions of Animal Viruses with Their Hosts
   A. In the case of animals, outcome of infection depends on many factors independent of the infected cell. These include defense mechanisms of the host.
   B. In most cases, viruses cause no obvious harm or disease.
   C. Most healthy humans carry a number of different viruses.
   D. Acute infections result in readily observed disease symptoms which disappear quickly.
   E. In persistent infections, the virus is present for many years without symptoms of disease.
        1. Chronic infections are the same as persistent infections.
        2. Slow viral diseases are a subgroup of persistent infections in which the disease progresses slowly and involves the central nervous system.
   F. Latent viral infections may result from integration of viral DNA into host cell DNA

V. Viruses and Tumors in Animals
   A. Tumors caused by DNA viruses result from integration and expression of viral DNA in the host cell
   B. Tumors caused by RNA viruses involve the making of a DNA copy of the viral RNA, which is then integrated into the chromosome of the host cell.

VI. Mechanisms of Cell Transformation by Viruses
   A. The integrated viral DNA disrupts control of normal cell growth and tissue development.
   B. Retroviruses integrate oncogenes, which are similar to normal host genes termed protooncogenes. Oncogene gene products probably disrupt the function of protooncogene gene products.

VII. Viruses and Tumors in Humans
   A. Three groups of DNA viruses have been shown to cause tumors in humans. These are the Epstein-Barr virus, hepatitis B virus, and the human papillomavirus.
   B. There are other possible cancer-causing viruses. These include polyoma virus and adenovirus.
   C. A rare human leukemia is caused by a retrovirus.
        1. Human T-cell lymphotropic virus causes a rare cancer of a specific white blood cell - the T-cell

VIII. HIV Disease AIDS
   A. HIV disease is caused by an RNA virus of the retrovirus family

IX. Viruses that Cause Disease of Plants
   A. A large number of serious plant diseases are caused by viruses.
   B. Plant viruses are spread by a variety of mechanisms.
        1. Plant viruses are very resistant to inactivation.
        2. Insects are most important for transfer.

X. Insect Viruses
   A. Some viruses can multiply in plants and insects.
   B. Some viruses are highly pathogenic for insects and are used in biological control of insects.

XI. Viroids
   A. Viroids, short single-stranded RNA molecules cause a variety of plant disease.

VOCABULARY: TERMS AND DEFINITIONS

_____ 1. Transforming genes of retroviruses which are integrated into the host chromosome

_____ 2. RNA viruses which make a DNA copy and incorporate it into the host cell genome

_____ 3. Protein surface projections from viruses that are used for attachment

_____ 4. A mechanism for releasing enveloped virions from a host cell

_____ 5. A term for a retrovirus after it has integrated into the host chromosome

_____ 6. Diseases of relatively short duration

_____ 7. A naked, small, single-stranded infectious RNA molecule

_____ 8. A method of viral penetration into animal cells

_____ 9. A hidden infection generally without symptoms

_____ 10. Individuals that harbor a virus, are infectious, but do not have a disease

_____ 11. A swelling that results from the abnormal growth of cells

_____ 12. Cells that have lost the ability to control their growth

a. spikes
b. endocytosis
c. budding
d. carriers
e. acute infections
f. retrovirus
g. latent infections
h. tumor
i. transformed cells
j. provirus
k. oncogenes
l. viroid

SELF-TEST OF READING MATERIAL

1. Animal viruses usually penetrate a host cell by
    a. injection.
    b. exocytosis.
    c. pili.
    d. endocytosis.
    e. a vector.

2. An envelope is acquired during which of the following steps?
    a. penetration
    b. lysis
    c. synthesis
    d. release
    e. assembly

3. When an animal virus becomes integrated into the host's chromosome it is now called a
    a. prophage
    b. provirus
    c. vegetative virus
    d. temperate virus
    e. temperate phage

4. Which of the following statements about spikes on viruses is true?
    a. They contain lysozyme for rupturing the host cell.
    b. They function in recognition and attachment to host cells.
    c. They are proteins.
    d. b and c are correct.
    e. a and b are correct.

5. An RNA virus with reverse transcriptase will first synthesize a
    a. mRNA molecule.
    b. complementary strand of DNA.
    c. cDNA molecule.
    d. complementary strand of RNA.
    e. rRNA molecule.

6. An example of a chronic virus infection is
    a. cancer.
    b. hepatitis.
    c. chicken pox.
    d. herpes.
    e. Two of the above are correct.

7. Virtually all slow viral infections involve
    a. the central nervous system.
    b. a viroid.
    c. retroviruses.
    d. plasmids.
    e. digestive system.

8. An example of a latent virus infection is
    a. cancer.
    b. hepatitis.
    c. SSPE
    d. herpes.
    e. Two of the above are correct.

REVIEW QUESTIONS

1. Distinguish between chronic viral infections and latent viral infections. Give examples of each.

2. Describe three mechanisms by which animal viruses enter the host cell.

3.Complete the following table.

NORMAL CELLS	CANCER (TRANSFORMED) CELLS
1. Cells grow as a monolayer.	1.
2. Cells attach to each other and glass surface.	2.
3.	3. Cells grow indefinitely.
4. Cells do not form tumors.	4.

4. Describe the four characteristics used to classify animal viruses.

5. Explain why bacteriophages often use lysozyme for penetration and release from host cells while animal and plant viruses do not.

ANSWERS:
   Vocabulary:Terms and Definitions
      1. k 2. f 3. a 4. c 5. j 6. e 7. l 8. b 9. g 10. d 11. h 12. i
   Self-Test of Reading Material
      1. d 2. d 3. b 4. d 5. d 6. b 7. a 8. d

15 NONSPECIFIC IMMUNITY

The body's defense against infectious disease takes two forms, nonspecific immunity that acts as barriers to infectious agents without regard to the specific organism, and specific immunity in which the responses depends on the specific organism. Mechanisms of nonspecific immunity include innate immunity that is not affected by prior contact, mechanical and chemical barriers that protect against infection, and inflammation. This chapter explores the various mechanisms involved in nonspecific immunity.

KEY CONCEPTS

1. Innate nonspecific immunity is not affected by prior contact with the infecting agent.
2. Tissue barriers and chemical factors contribute to nonspecific immunity.
3. Inflammation and physiological changes are also important in nonspecific immunity.

SUMMARY

I. The Host Defends Itself: Nonspecific and Specific Defense Mechanisms
    A. Nonspecific mechanisms include physical barriers as well as physiological mechanisms such as fever and inflammation.
    B. Specific mechanisms include specialized cells and proteins that show an enhanced response to repeat infections.

II. Cells and Tissues Involved in Immune Responses
    A. Cells of the blood and lymphoid system participate in immune responses.
        1. Leukocytes are the cells primarily responsible for defense.
            a. Granulocytes are leukocytes that contain distinguishing granules. They include neutrophils, eosinophils, and basophils.
            b. Mononuclear phagocytes such as the macrophages and monocytes, do not contain granules.
            c. Lymphocytes are leukocytes that are involved in the specific immune response.

III. Innate Nonspecific Immunity
    A. Tissue barriers and nonspecific factors are important in nonspecific immunity.
    B. Physical barriers include an intact skin and mucus membranes that line body cavities.
    C. Nonspecific antimicrobial factors include the pH of body fluids and surfaces, and factors such as lysozyme, beta-lysin, and peroxidase enzymes.

IV. Complement is of Primary Importance as an Antimicrobial Factor and in Inflammation
    A. Complement consists of at least 26 different proteins that act in sequence one after another in a cascade fashion.
    B. Complement can be activated in two ways.
        1. The classical pathway requires a specific immune reaction for activation.
        2. The alternative pathway does not require a specific immune reaction for activation.

V. Cytokines Are Also Nonspecific Factors Important in Defense.
    A. Cytokines are molecular messages essential for communication between immunologically active cells.
    B. The groups of cytokines include three important factors.
        1. Interferons are antiviral glycoproteins.
        2. Interleukins function in many different ways.
        3. Colony stimulating factors are important in the multiplication and differentiation of blood cells.

VI. Inflammation is an Early Reaction to Injury.
   A. The principle events of inflammation include:
      1. Tissue injury followed by the release of chemical mediators.
      2. Blood vessel dilation and the increased flow of plasma together with redness, pain, and heat.
      3. White blood cell migration to the area of injury.

VII. Phagocytosis is Essential to Nonspecific Defense.
   A. Chemical products of microorganisms, components of complement, and products from injured cells act as chemoattractants to phagocytes.
   B. After engulfment, microorganisms are contained in a phagolysosome.
   C. Microorganisms are killed and digested within the phagolysosome.

VIII. Physiological Change Affects the Immune Response
   A Fever adversely affects the growth of microorganisms and enhances host defense mechanisms including inflammation and phagocyte killing.
   B. Iron is bound in the body by transferrin and lactoferrin.
   C. Protein and carbohydrate metabolism increase in response to infection.

VOCABULARY:TERMS AND DEFINITIONS

The following list contains new terms introduced in this chapter. Use these terms to fill-in the blanks of the sentences that follow and you will have a definition or description of each new term.

inflammation   leukocytes   macrophages
granulocytes   giant cells   lysozyme
complement   cytokines   interferons
phagolysosome   interleukin-1

1. In response to an infection such as tuberculosis, a number of macrophages may fuse together to form _____ .

2. _____ are molecular messages essential for communication between immunologically active cells.

3. _____ is an enzyme found in body fluids that attacks the peptidoglycan layer of bacterial cell walls.

4. Antiviral glycoproteins that interfere with virus replication are called _____ .

5. Another name for white blood cells is _____ .

6. A _____ is the result of the fusion of a phagosome and a lysosome in a phagocytic cell.

7. Along with the monocytes, the _____ comprise a widespread system of cells known as the mononuclear phagocyte system.

8. _____ is a nonspecific reaction to tissue injury.

9. The cytokine, _____ , is released from phagocytic cells and acts as a message to the hypothalamus that microorganisms have invaded the body.

10. _____ is a group of at least 26 different proteins that act in sequence one after another in a cascade fashion.

11. _____ are white blood cells that contain cytoplasmic granules with distinguishing staining characteristics.

SELF-TEST OF READING MATERIAL

1. Interleukin-1 is a protein produced by invading bacteria which is carried to the hypothalamus by the bloodstream and the hypothalamus responds by increasing the body temperature.
   a. True
   b. False

2. Complement is a group of interacting serum proteins which provides a nonspecific host defense mechanism.
   a. True
   b. False

3. Vacuoles within macrophages which contain a variety of digestive enzymes are called
   a. lysosomes.
   b. ligands.
   c. lysozymes.
   d. lister vesicles.

4. Which of the following blood cells are the first to appear in response to an infection?
   a. eosinophils
   b. erythrocytes
   c. lymphocytes
   d. basophils
   e. neutrophils

5. Which of the following nonspecific substances is a normal component of serum?
   a. complement
   b. lysozyme
   c. interferon
   d. collagenase
   e. lysosome

6. After invading bacteria are engulfed by professional phagocytes, they are contained within a vacuole called a
   a. lysosomes.
   b. phagosome.
   c. lysozymes.
   d. pinosome.

7. Which of the following are nonspecific means of defense against foreign invaders?
   a. intact skin
   b. flushing action of urinary tract
   c. inflammation
   d. fever
   e. All of the above are nonspecific means of defense.

8. Specific immune responses are most highly developed in
   a. mammals.
   b. birds.
   c. reptiles.
   d. a and b are correct.
   e. a and c are correct.

9. During phagocytosis, when a phagosome containing foreign invading bacteria fuses with a lysosome, the resulting vacuole is called a
   a. lysosomes.
   b. phagosome.
   c. lysozymes.
   d. phagolysosome

10. The observation that humans are not infected with cat distemper virus can be partly explained because of
   a. differences in the virus receptors.
   b. differences in host cell receptors.
   c. age differences.
   d. differences in host cell ligands.

11. Which of the following are professional phagocytes?
    a. neutrophils
    b. monocytes
    c. macrophages
    d. 1 and 3 only are correct.
    e. 1, 2, and 3 are correct.

12. Where are phagocytes found in the body?
    1. blood
    2. liver
    3. spleen
    4. mucous membranes
    5. lungs
    6. nasal passages
    a. 1,3,5
    b. 1,2,4,5
    c. 1 only
    d. 1,2,3,4,5,6
    e. 1 and 5 only

13. During killing in a phagolysosome, oxygen consumption
    a. greatly increases.
    b. remains essentially unchanged.
    c. greatly decreases.

14. All blood cells including leukocytes, erythrocytes and platelets originate from
    a. spleen cells.
    b. the lymphatic system.
    c. liver cells.
    d. hematopoietic stem cells.

15. Which of the following blood cells account for more than half of the leukocytes in human blood?
    a. neutrophils
    b. basophils
    c. eosinophils
    d. macrophages

16. Which of the following blood cells typically increase in numbers during an active infection?
    a. neutrophils
    b. basophils
    c. eosinophils
    d. macrophages

REVIEW QUESTIONS

1. The treatment for chronic, active hepatitis B, a viral disease of humans, is interferon. Can you explain why?

2. Outline the principle events occurring during inflammation.

3. Complete the following table of types of human leukocytes.

CELL TYPE	PERCENTAGE IN BLOOD	FUNCTION
1. 2. 3. basophils	55-65% 2-4%	
MONONUCLEAR PHAGOCYTES 1. 2.		Phagocytize and digest engulfed materials and play an important role in development of specific immunity
LYMPHOCYTES	25-35%	

4. What is opsonization and how does it occur?

ANSWERS
Vocabulary:Terms and Definitions
1. giant cells 2. cytokines 3. lysosome 4. interferons 5. leukocytes 6. phagolysosome 7. macrophages 8. inflammation 9. interleukin-1 10. complement 11. granulocytes
Self-Test of Reading Material
1. b 2. a 3. a 4. e 5. a 6. b 7. e 8. d 9. d 10. b 11. e 12. d 13. a 14. d 15. a 16. a

NOTES

16 SPECIFIC IMMUNITY

The immune response is characterized by its specificity and a memory response. The immune response requires the coordination of a number of cellular and molecular events to provide protection. It is difficult to find a starting place because the events are so closely related. Many aspects of the immune response still remain a mystery, but from what is known, it is an amazing and beautiful process.

KEY CONCEPTS

1. The specific immune response explains why immunity to one disease does not usually confer immunity to another.
2. The specific immune response depends upon two types of lymphocytes interacting with each other and with other cells.
3. The specific immune response has "memory" which enhances the response and rapidly eliminates foreign substances when they are reintroduced into the body. Vaccination takes advantage of this phenomenon.
4. The specific immune response can discriminate between self and nonself.

SUMMARY

I. Specific immune response has three features:
    A. Response is highly specific.
    B. Response has memory.
    C. Two sets of lymphocytes involved, the B and the T cells.

II. The Nature of Antigens
    A. Antigens are usually foreign to the host.
    B. Antigens are usually macromolecules.
    C. The immune response is directed to antigenic determinants or epitopes.
    D. Haptens must bind to larger antigenic molecules to induce an immune response.

III. The Nature of Antibodies
    A. Antibody molecules are proteins known as immunoglobulins.
    B. Antibody molecules are Y-shaped with antigen reaction sites at the ends of the arms of the Y.
    C. Antigen-antibody binding depends on a close complementary fit.
    D. Immunoglobulin classes have different properties.
        1. IgG is the most abundant and the only one that crosses the placenta to protect the fetus and newborn.
        2. IgM is the first antibody to form and is involved in complement fixation.
        3. IgA is found on mucous surfaces
        4. IgD functions in development and maturation of antibody response.
        5. IgE attaches to mast cells and basophils and participates in allergic reactions.

IV. The Role of Lymphocytes in Specific Immunity
    A. Lymphocytes are the cells primarily responsible for the specific immune response.
    B. There are two types of lymphocytes, the T lymphocytes and the B lymphocytes.
    C. Markers on the surface of the lymphocytes distinguish various subgroups.
    D. Most T lymphocytes are CD4$^+$ helper T cells or CD8 cytotoxic T cells.
    E. B lymphocytes become plasma cells and synthesize specific antibody.
    F. Major histocompatibility complex antigens are recognition molecules on cell surfaces.

V. Development of the Antibody Response
    A. Thymus-dependent antigens must be processed by cells, such as macrophages, and presented to T cells in special ways before the T cells can stimulate the B cells to synthesize antibody.
    B. Thymus-independent antigens, such as certain bacterial lipopolysaccharides and polysaccharides, can stimulate B cells directly.
    C. Clonal selection occurs during immune responses.
    D. The memory response is faster and more efficient than the primary response.
    E. Immune specificity permits millions of different antigens to be recognized.

VI. Development of the Cell-Mediated Immune Response
    A. The cell-mediated immune response involves T lymphocytes.
    B. Cytotoxic T cells kill virus-infected cells and other cells recognized as foreign.
    C. Immunity to most cancers and to tissue transplants is cell-mediated.
    D. Cell-mediated immunity controls microbial growth in macrophages.
    E. Natural killer cells kill foreign cells, but are not antigen-specific.

VII. Immunological Tolerance and Control
    A. Immunological tolerance denotes specific unresponsiveness to antigen.
    B. The immune response is carefully controlled.

VOCABULARY: TERMS AND DEFINITIONS

The following list contains new terms introduced in this chapter. Use these terms to fill-in the blanks of the sentences that follow and you will have a definition or description of each new term.

B cells	**antigenic determinants**	**T cells**
haptens	**antibodies**	**clone**
opsonins	**antigens**	**immunoglobulins**
anamnestic		

1. _____ molecules are proteins known as immunoglobulins.

2. A _____ is a group of cells derived from a single cell.

3. _____ are lymphocytes that differentiate into antibody-producing plasma cells on antigenic stimulation.

4. _____ are antibodies that coat an antigen and make it more susceptible to phagocytosis.

5. Specific chemical groups on antigens toward which the immune response is directed are

   called _____ .

6. Certain kinds of substances that are foreign to the host and are able to trigger an immune response in the host are

   called _____ .

7. _____ are lymphocytes that do not secrete antibodies but instead act directly on other cells.

8. Antibodies belong to the class of serum proteins called _____ .

9. _____ are small molecular weight molecules that must combine with larger molecules such as proteins in order to incite an immune response.

10. The memory response is also known as the _____ response.

SELF-TEST OF READING MATERIAL

1. Which of the following molecules would be the best antigens?
    a. lipids
    b. glucose
    c. fatty acids
    d. proteins
    e. sugars

2. Which part of the antibody molecule reacts with the antigenic determinants of the antigen?
    a. hinge
    b. H chains
    c. constant sections of the H and L chains
    d. L chains
    e. variable sections of the H and L chains

3. An antigen is
    a. a hapten that combines with an antibody.
    b. a substance that incites an antibody response and can combine specifically with these antibodies.
    c. a small molecule that attaches to cells.
    d. a carbohydrate.
    e. a protein that combines with antibodies.

4. Which of the following best describes some characteristic of B cells?
    a. They differentiate into antibody-producing cells.
    b. They originate in the gut.
    c. They differentiate into neutrophils.
    d. They differentiate and mature in the thymus.
    e. They function in cell mediated immunity.

5. Which of the following best describes some characteristic of T cells?
    a. They differentiate into antibody-producing cells.
    b. They originate in the gut.
    c. They differentiate into neutrophils.
    d. They differentiate into macrophages.
    e. They function in cell mediated immunity.

6. The rapid increase and production over a longer period of antibodies in response to a second or subsequent exposure to an antigen is known as
    a. a primary response.
    b. the killer response.
    c. a memory or anamnestic response.
    d. inflammation.
    e. dose response.

7. An immunoglobulin is a(n)
    a. histamine.
    b. antigen.
    c. macrophage.
    d. antibody.
    e. carbohydrate.

8. The portion of an antigen molecule that stimulates an antibody response is called the
    a. antigenic determinant.
    b. epitope.
    c. antibody determinant.
    d. a and b are correct.
    e. b and c are correct.

9. Which of the following classes of immunoglobulins can cross the placenta?
    a. IgG
    b. IgM
    c. IgA
    d. IgD
    e. IgE

10. Macrophages play an important role in the immune response because they present a "processed" antigen to lympho-cytes.
    a. True
    b. False

11. Autoimmune diseases result because the immune system fails to recognize host cells foreign.
    a. True
    b. False

12. Which of the following classes of immunoglobulins are described as secretory?
    a. IgG
    b. IgM
    c. IgA
    d. IgD
    e. IgE

13. Plasma cells are
    a. mature T cells.
    b. immature macrophages.
    c. antibody-producing cells.
    d. mature macrophages.
    e. immature T cells.

14. B lymphocytes originate in
    a. adult bone marrow.
    b. thymus.
    c. fetal liver.
    d. a and b are correct.
    e. a and c are correct.

15. Antibody molecules are
    a. carbohydrates.
    b. lipids.
    c. carbohydrates.
    d. proteins.
    e. peptidoglycans.

REVIEW QUESTIONS

1. Distinguish among resistance, innate resistance, and acquired immunity.

2. List the types of immunoglobulins along with a description of their functions.

3. Describe the types of T cells along with their functions.

4. Outline the major steps in an immune response to a thymus-dependent antigen.

ANSWERS:
   Vocabulary: Terms and Definitions
      1. antibodies 2. clone 3. B cells 4. opsonins 5. antigenic determinants 6. antigens 7. T cells 8. immunoglobulins
      9. haptens 10. anamnestic
   Self-Test of Reading Material
      1. d 2. e 3. b 4. a 5. e 6. c 7. d 8. d 9. a 10. a 11. b 12. c 13. c 14. e 15. d

---

NOTES

17 FUNCTIONS AND APPLICATIONS OF IMMUNE RESPONSES

The previous chapters described the mechanisms of nonspecific and specific immunity. This chapter explores how antibodies and immune T cells actually protect against infection and disease. After all of the synthesis of antibodies and production of immune cells, it will be helpful to see how these things work in immunity. Antigen and antibody reactions *in vitro* serve as the basis for many serological laboratory tests. The intentional introduction of killed or attenuated infectious agents or other substances for the purpose of stimulating an immune response is the basis of vaccination. Vaccines as immunizing agents are being improved with new technology.

KEY CONCEPTS

1. Antibodies and immune T cells protect the human body by diverse mechanisms.
2. *In vitro* immunological reactions are the basis of many laboratory tests.
3. Immunization, or vaccination, can control or even eliminate some (but not all) serious infectious diseases.
4. Immunizing agents have been dramatically improved with new technology.

SUMMARY

I. Functions of Antibodies *in vivo*
    A. Antibodies function in a number of ways:
        1. Precipitins cause antigens to precipitate while agglutinins cause antigens to agglutinate. Both functions permit more efficient phagocytosis.
        2. Opsonins coat an antigen so it can be more readily phagocytized.
        3. Complement-fixing antibodies initiate the complement cascade and cause complement lysis of the antigen.
        4. Antitoxins neutralize the effects of toxins while neutralizing antibodies prevent viruses from infecting cells.
    B. IgM antibodies are the first antibodies synthesized in response to an infection and are very effective agglutinins. They also are effective at activating complement.
    C. IgG antibodies are effective agglutinins, precipitins, opsonins and neutralizing antibodies.
    D. IgA antibodies inhibit the adherence of microorganisms to human cells.
    E. IgE antibodies protect against certain parasitic worm infections and participate in the allergic response.

II. Functions of Cell-Mediated Immunity *in vivo*
    A. Immune cytotoxic T cells kill virus-infected cells, tumor cells, and tissue transplants.
    B. Membrane-bound antigens serve as targets for cytotoxic T cells.
    C. Activation of cells such as macrophages enhances their ability to find and kill infectious organisms such as *Mycobacterium tuberculosis*.

III. Applications of Immune Responses *in vitro*
    A. Serology is the branch of immunology that applies *in vivo* functions *in vitro*.
    B. Precipitation reactions involving soluble antigens, form the basis for many immunological tests.
        1. Precipitation occurs in a zone of optimal proportion.
        2. The Ouchterlony technique is a precipitation test in agar by immunodiffusion.
        3. Immunoelectrophoresis combines immunodiffusion in a gel with electrophoresis.
    C. Agglutination reactions involve particulate antigens.
        1. Direct and indirect agglutination tests can give an estimation of the antibody titer, that is the amount of antibody present in serum.
        2. Hemagglutination tests use red blood cells as indicators.

D. Virus neutralization test for virus infections.
E. Immunofluorescence tests use a visible tag to see the antigen or antibody.
F. Complement fixation tests involve a specific antibody acting with complement to cause cell lysis.
G. Radioimmunoassay (RIA) and enzyme linked immunosorbent assay (ELISA) use radioactive labels or enzyme reactions to visualize antigen-antibody reactions.
H. The Western blot or immunoblot combines ELISA and electrophoresis to separate antigens.

IV. Vaccination and Immunization
A. Immunity can be active or passive and natural or artificial..
B. Passive immunity usually involves the transfer of preformed antibodies.
C. Active immunity usually involves the synthesis of antibodies by the host.
D. Active immunity is induced artificially by living, killed or toxoid vaccines.
E. New technology has produced subunit vaccines, recombinant vaccines, and peptide vaccines.
F. The importance of routine immunizations for children cannot be overemphasized.

VOCABULARY: TERMS AND DEFINITIONS

The following list contains new terms introduced in this chapter. Use these terms to fill-in the blanks of the sentences that follow and you will have a definition or description of each new term.

vaccination   toxoids   titer
active immunity   attenuated   natural immunity
immune serum globulin passive immunity precipitins
artificial immunity   agglutinins   serology

1. A measure of the amount of antibody in serum is known as the antibody _____ .

2. _____ usually involves the transfer of preformed antibodies produced by other people or animals.

3. Immunity resulting from the deliberate exposure to antigens for the purpose of protection is

   called _____ .

4. Microorganisms and viruses, although viable, that have been modified so as to be incapable of causing disease under

   ordinary circumstances are said to be _____ .

5. _____ are relatively large, particulate antigens.

6. Molecular or soluble antigens are known as _____ .

7. _____ is a term derived from the Latin word meaning cow and is used to describe a protective inoculation.

8. _____ , or gamma globulin, is a pooled plasma fraction from donors that contains antibodies to certain diseases.

9. _____ is immunity produced by the individual in response to an antigenic stimulus.

10. Exposure of individuals to infectious agents or other immunizing agents that induce an immune response results

    in _____ .

11. The branch of immunology that uses in vitro antigen-antibody reactions in laboratory tests

    is called _____ .

12. _____ are bacterial toxins treated to destroy the toxic part of the molecule while retaining the antigenic part and are used for immunization.

SELF-TEST OF READING MATERIAL

1. Infectious organisms that are, although still alive, incapable of causing disease are said to be
   a. virulent.
   b. pathogenic.
   c. attenuated.
   d. Such organisms do not exist.

2. A toxoid is a(n)
   a. type of antibody that combines with a toxin.
   b. type of enzyme that destroys toxins.
   c. inactivated toxin.
   d. type of bacterium that resists phagocytosis.
   e. type of virus.

3. Which of the following is the type of immunity produced after recovery from a disease?
   a. innate
   b. naturally acquired active
   c. naturally acquired passive
   d. artificially acquired active
   e. artificially acquired passive

4. Immunity of humans to cat and dog distemper is which of the following?
   a. innate
   b. naturally acquired active
   c. naturally acquired passive
   d. artificially acquired active
   e. artificially acquired passive

5. Administration of gamma globulin for hepatitis would produce which type of immunity?
   a. innate
   b. naturally acquired active
   c. naturally acquired passive
   d. artificially acquired active
   e. artificially acquired passive

6. Transfer of immunity from mother to fetus across the placenta
   a. innate
   b. naturally acquired active
   c. naturally acquired passive
   d. artificially acquired active
   e. artificially acquired passive

7. Immunity after poliovirus vaccine
   a. innate
   b. naturally acquired active
   c. naturally acquired passive
   d. artificially acquired active
   e. artificially acquired passive

8. Which of the following types of immunity is independent of previous exposure to foreign substances?
   a. innate
   b. naturally acquired active
   c. naturally acquired passive
   d. artificially acquired active
   e. artificially acquired passive

9. Which of the following types of immunity is not very long lasting?
    a. naturally acquired active
    b. artificially acquired passive
    c. naturally acquired passive
    d. artificially acquired active
    e. b and c are correct.

Match the serological test on the right with its description on the left.

10. Antibodies are mixed with particulate antigens such as bacteria, fungi or red blood cells.
11. Serum is mixed with known viral suspensions before the virus is used to infect a cell culture.
12. Antigen and antibody are allowed to diffuse through agar to form distinctive lines of precipitate.
13. This test uses and enzyme reaction such as peroxidase to label an antigen-antibody reaction.
14. Antigens are added to red blood cells or latex beads before reacting with an antibody.
15. This technique combines ELISA with electrophoresis to separate antigens.
16. Test organisms are fixed to a slide before reacting with a fluorescein labeled antibody.
17. Antibodies to viruses are detected by the inhibition of clumping of red blood cells.
18. Radioactively labeled material reacts with antibody bound to antigen.
19. This test combines precipitation with electrophoresis.
20. Fluorescent labeled anti-human-gamma-globulin is reacted with antigen-antibody complexes.

a. Ouchterlony technique
b. immunoelectrophoresis
c. direct agglutination
d. indirect agglutination
e. hemagglutination
f. neutralization
g. direct fluorescence
h. indirect fluorescence
i. RIA
j. ELISA
k. Western blot

REVIEW QUESTIONS

1. Distinguish among subunit vaccines, recombinant vaccines, and peptide vaccines.

2. What are some advantages of the new technology vaccines?

3. Describe for each class of antibody, its ability to protect against infection and disease.

4. How do T cells protect against virus-infected cells and tumors?

5. Why is it not possible to vaccinate against every infectious agent that threatens our health?

NOTES

ANSWERS

Vocabulary: Terms and Definitions

1. titer 2. passive immunity 3. artificial immunity 4. attenuated 5. agglutinins 6. precipitins 7. vaccination 8. immune serum globulin 9. active immunity 10. natural immunity 11. serology 12. toxoids

Self-Test of Reading Material

1. c 2. c 3. b 4. a 5. e 6. c 7. d 8. a 9. e 10. c 11. f 12. a 13. j 14. d 15. k 16. g 17. e 18. i 19. b 20. h

18 IMMUNOLOGIC DISORDERS

A healthy, effectively functioning immune system is a beautiful thing. Its beauty is because it is a very complicated thing that requires the coordination of many individual parts to protect against infection. Sometimes portions of the system fail and the results can be more detrimental than useful. When the system fails to respond adequately to antigenic stimulation, an immunodeficiency occurs. This failure may be an inherited defect or acquired as in AIDS. At times our immune system appears to overreact to antigenic stimulation and instead of protecting us against certain antigens, it actually causes damage to us. This is the situation known as hypersensitivity or allergies. Finally, in autoimmunity, our immune systems may fail to recognize our own body antigen as self and begin to produce antibodies against our own tissue antigens. All three conditions can have very serious and sometimes devastating effects.

KEY CONCEPTS

1. Hypersensitivities result from the immune system reacting inappropriately to certain antigens.
2. Immunodeficiency results when the immune system cannot respond adequately to antigenic stimulation: overproduction of immune factors can also occur.
3. Autoimmune disease results when the immune system responds to tissues in its own body.

SUMMARY

I. Hypersensitivities
    A. Type I or immediate hypersensitivities are IgE mediated.
        1. IgE antibodies attach to receptors on mast cells and basophils.
        2. When antigen combines with these cell-fixed IgE antibodies, histamine, and other chemical mediators are released.
        3. Localized anaphylaxis such as hives hayfever, and asthma are caused by these chemical mediators..
        4. Generalized anaphylaxis is rare but can occur after a bee sting or penicillin reaction.
        5. Immunotherapy involves repeatedly injecting a small but increasing amount of the offending antigen in order to elicit an IgG response.
    B. Type II hypersensitivity or cytotoxic reactions are caused by humoral antibodies that can destroy normal cells.
        1. Transfusion reactions
            a. ABO blood group system
            b. ABO antigens are polysaccharides on the red cells.
        2. Hemolytic disease of the newborn
            a. Rh or Rhesus blood group system is involved.
            b. If an Rh negative mother carries an Rh positive fetus, the Rh positive cells cause antibody production in the mother. These antibodies can cross the placenta and destroy fetal red cells.
            c. A new treatment involves injecting anti-Rh antibodies into the mother shortly after delivery or abortions to develop anti Rh antibodies.
    C. Type III hypersensitivity or immune complex reactions activate inflammatory mechanisms
        1. Immune complexes have considerable biological activity such as blood clotting mechanisms, attraction of leukocytes, and stimulation of proteolytic enzyme release.
        2. Immune complexes can be deposited in kidneys, joints, and skin.
        3. The arthus reaction causes the death of tissue following injection of antigen into previously immunized animal.
        4. Other diseases, such as serum sickness and farmer's lung, are mediated by immune complexes.
    D. Type IV Hypersensitivity or delayed hypersensitivity involves T lymphocytes.
        1. The response is delayed; developing over several days.
        2. Tuberculin skin test is an example of delayed hypersensitivity.
        3. Contact allergies are reactions to metals, poison ivy and similar products.

        4. Delayed hypersensitivity in infectious diseases
           a. Leprosy and hepatitis are examples.
           b. Cell destruction with impairment of tissue function is involved.
    E. Transplantation
        1. MHC matching essential to avoid rejection.
        2. The graft versus host disease is a delayed type reaction.
        3. Use of immunosuppressant drugs such as cyclosporin has helped keep graft rejection under control.

II. Immunodeficiency Disorders
    A. Primary immunodeficiencies result from genetic or developmental abnormalities.
        1. In DiGeorge's syndrome, the thymus fails to develop and patient lacks T lymphocytes.
        2. X-linked agammaglobulinemia patients lack lymphocytes.
        3. In severe combined immunodeficiency disease the bone marrow stem cells are defective.
        4. Chediak-Higashi disease affects the phagocytes.
    B. Secondary immunodeficiency results from malnutrition, immunosuppressive agents, infections, and some malignancies.
        1. Multiple myelomas are cancers.
        2. Acquired immune deficiency disease (AIDS)

III. Autoimmune Diseases
    A. Some autoimmune diseases are caused by the production of antibodies to structures within the body: myasthenia gravis
    B. Some autoimmune diseases are mediated by mechanisms of delayed hypersensitivity: sympathetic ophthalmia.

SELF-TEST OF READING MATERIAL

1. An exaggerated or inappropriate immune response is known as a(n)
    a. immunodeficiency.
    b. precipitation.
    c. histamine.
    d. hypersensitivity.
    e. allergen.

2. Localized anaphylaxis involving the skin is called
    a. hay fever.
    b. asthma.
    c. shock.
    d. hives.
    e. hayfever.

3. Immunotherapy to prevent generalized anaphylaxis is done by injecting dilute doses of
    a. IgG antibodies.
    b. antihistamine.
    c. IgE antibodies.
    d. offending antigen.
    e. offending antibody.

4. Which of the following is most commonly involved in graft rejections?
    a. ABO antigens
    b. ABO antibodies
    c. MHC antigens
    d. MHC antibodies
    e. a and c are correct.

5. Which of the following would be a Type IV, or delayed hypersensitivity?
    a. hayfever
    b. allergic contact dermatitis
    c. glomerulonephritis
    d. penicillin reaction
    e. blood transfusion reaction

6. The symptoms of an immune complex reaction are due to
    a. an inflammatory response.
    b. histamines.
    c. autoimmune antigens.
    d. IgE.
    e. autoimmune antibodies.

7. Autoimmunity is a delayed hypersensitivity or Type IV reaction.
    a. True
    b. False

8. Allergic contact dermatitis is mediated by
    a. sensitized T cells.
    b. IgG antibodies.
    c. IgE antibodies.
    d. basophils and mast cells.
    e. sensitized macrophages.

9. A hypersensitive reaction occurs
    a. during the first exposure to an antigen.
    b. in individuals with diseases of the immune system.
    c. on a second or subsequent exposure to an antigen.
    d. during autoimmune diseases.
    e. in immunologically deficient individuals.

10. Type I hypersensitivities
    1. are cell mediated.
    2. are IgE mediated.
    3. have many symptoms which are due to histamine.
    4. are caused by antibodies bound to basophils and mast cells.
    5. have symptoms which occur soon after exposure to the allergen.
    a. 1,4,5
    b. 2,3,4,5
    c. 1
    d. 2,4,5
    e. 1,3,4,5

11. Hemolytic disease of the newborn occurs when an
    a. Rh-positive mother carries an Rh-negative fetus.
    b. O mother carries an AB fetus.
    c. Rh-negative mother carries an Rh-positive fetus
    d. AB mother carries an O fetus.
    e. Two of the above are correct.

12. Reaction of antigen with IgE antibodies attached to mast cells causes
    a. precipitation.
    b. complement fixation.
    c. degranulation.
    d. agglutination.
    e. asthma.

13. Immediate hypersensitivities are mediated by
    a. allergens.
    b. macrophages.
    c. humoral antibodies
    d. antigens.
    e. T cells.

14. Immunotherapy for hypersensitivities probably works because of
    a. the production of IgG blocking antibodies.
    b. development of suppressor T cells.
    c. saturation of IgE antibodies.
    d. All of the above are correct.
    e. a and b are correct.

15. Delayed hypersensitivities are mediated by
    a. allergens
    b. macrophages
    c. humoral antibodies
    d. antigens
    e. T cells

16. Farmer's lung is the result of
    a. sensitized T cells.
    b. immune complex reactions.
    c. Type I hypersensitivity.
    d. cytotoxic reactions.
    e. urticaria

17. Autoimmune diseases result when the immune response recognizes host cells as antigenic.
    a. True
    b. False

REVIEW QUESTIONS

1. Distinguish between primary or congenital and secondary or acquired immunodeficiency disease.

2. What conditions can cause secondary or acquired immunodeficiency disease.

3. Give an example of an immunodeficiency disease that involves each of the following:

   a. B cell deficiency

   b. T cell deficiency

   c. B and T cell deficiency

   d. defective phagocyte

4. Describe for each of the above conditions what would be missing in the immune response.

5. Describe the events of degranulation.

6. Why don't antihistamines work for asthma?

NOTES

19 INTERACTIONS BETWEEN MICROORGANISMS AND HUMANS

When a microorganism infects a human host, the outcome of that infection will depend on a dynamic interaction between the microorganism and the host. When the host defenses adequately protect the host, the infection is resolved, but when the infecting microorganism gets the upper hand, disease occurs. Good health, free of infection, is the result of this dynamic balance and depends on several factors. These include the person's defenses to resist infection and disease and the factors of the microorganism that help it overcome host defenses. This chapter explores the nonspecific defense mechanisms of people and the virulence factors that microbes use to counter these defense mechanisms.

KEY CONCEPTS

1. Establishing and maintaining the normal flora as a dynamic human-microbe system is important for good health.
2. Microorganisms communicate with their human hosts and force the host cells to do things they would not normally do.
3. Pathogens have a variety of virulence factors responsible for their ability to cause disease. These factors are synthesized only when the bacteria need them.
4. The outcome of an infection depends on the host response as well as the activities of the infecting agent.

SUMMARY

I. Koch's postulates help establish the relationship between microorganisms and infectious disease.
    A. A molecular version of Koch's postulates is currently coming into use.

II. Normal flora of the human body.
    A. Microorganisms and humans form a variety of symbiotic relationships.
        1. Normal microbial flora constitute those organisms that colonize the host but do not normally produce disease.
        2. Commensalism is an association in which one partner benefits but the other remains unaffected.
        3. Mutualism refers to an association in which both partners benefit.
        4. Parasitism refers to an association in which one organism benefits at the expense of the other organism.
    B. Establishing and maintaining the normal flora as a dynamic human-microbe ecosystem is important for good health.
        1. Normal flora is acquired from contacts with the environment.
        2. Normal flora is important to the health of the host.
            a. Normal flora excludes potentially harmful organisms.
            b. Normal flora competes for nutrients and thus may exclude harmful organisms.
            c. Normal flora stimulates the immune system.
    C. The human-microorganism ecosystem is continually undergoing changes as a result of external factors including food, antibiotics, and moisture.

III. Infectious Disease Terminology
    A. Avirulent refers to microorganisms that lack the ability to cause disease.
    B. Colonization implies the establishment of microbial growth on a body surface.
    C. Disease describes any condition in which there is a noticeable impairment of body function.
    D. Infection occurs when a microorganism penetrates the body surfaces, enters the body tissue, multiplies and causes the host to react with an immune or other type of response.
    E. Infectious disease describes any condition in which there is a noticeable impairment of body function caused by an infecting microorganism or virus.
    F. An opportunist is a microorganism that is able to produce disease only in hosts with impaired defense mechanisms.

G. A pathogen is any disease-causing microorganism that commonly causes infection.

H. Pathogenic means disease-causing.

I. Virulence is a measure of pathogenicity and describes attributes of a microorganism or virus that promote its pathogenicity.

IV. Mechanisms of pathogenesis

A. Infectious microorganisms can come from a variety of sources including other humans, soil, air and water.

B. Attachment is usually a necessary step in the establishment of an infection.

1. Attachment by a microorganism does not necessarily imply infection, but is often a necessary prerequisite.

2. The ligands of the microorganism fit in a lock-key arrangement with a receptor on the cell surface.

3. Different microbial strains have different ligands; different cells have different receptors.

C. The establishment of infection usually involves the invasion of host tissues.

1. Some epithelial cells ingest microorganism by a process resembling phagocytosis.

2. Large concentrations of microorganisms may permit tissue invasion of the host tissues.

D. Colonization is a necessary step to establishing an infection.

E. Modes of interaction of microorganism with their hosts:

1. Pathogens can be described as extracellular, facultatively intracellular, and obligately intracellular.

F. Pathogens can interfere with phagocytosis and thereby establish infection.

1. Phagocytic killing involves the attracting the phagocyte to the microorganism, engulfing the microorganism and killing the microorganism.

2. Some phagocytes can avoid phagocytosis because they have a large capsule that prevents engulfment.

3. Some microorganisms can remain alive and even multiply within the phagocytes.

V. Toxins and Other Extracellular Products May Be Responsible For the Ability of Microorganisms to Cause Disease

A. Exotoxins are highly toxic, distinctive, heat-labile proteins released by certain species of microorganisms.

B. Endotoxins are heat-stable lipopolysaccharides from the cell walls of Gram-negative bacteria which play a contributing role in infectious disease.

VI. Evasion of the Host's Immune Response

A. Exit from the host and survival in the environment are essential steps in the infection process.

B. Impaired body defenses promote infectious disease.

C. Genetic factors, age, and stress are important in determining the susceptibility of people to certain infections.

D. Some potential pathogens live in host tissues for years held in check by host defenses.

VOCABULARY: TERMS AND DEFINITIONS

The following list contains new terms introduced in this chapter. Use these terms to fill-in the blanks of the sentences that follow and you will have a definition or description of each new term.

symbiosis	virulence	lipid A
pathogen	receptors	disease
normal flora	ligands	transient flora
opportunists	infection	

1. Host surface cell components which bind specifically to the surface of microorganisms are known

   as _____.

2. _____ describes interactions that occur between different organisms who live in close association with each other on a more or less permanent basis.

3. The condition in which there is a noticeable impairment of body function is called _____.

4. The microorganisms that grow on the external and internal surfaces of the body without producing obvious harmful

    effects to these tissues are known as the _____ .

5. _____ are members of the normal flora that are able to cause disease under special circum-
    stances.

6. Properties of microorganisms that enhance their disease producing capabilities contribute to their_____ .

7. _____ occurs when a microorganisms penetrates the body surfaces, enters the body tissue, multiplies and
    causes the host to interact with an immune or other type of response.

8. Microorganisms that inhabit the body only occasionally are termed _____ .

9. _____ is the lipid component of endotoxins that is responsible for the toxic properties of the
    endotoxin.

10. Projections from the surface of microorganisms or viruses which bind specifically to host cells

    are called _____ .

11. A _____ is any disease-producing microorganism.

SELF-TEST OF READING MATERIAL

1. Which of the following may contribute to a pathogen's virulence?
   1. endotoxins
   2. capsules
   3. ligands
   4. collagenase
   5. exotoxins

   a. 2,5
   b. 1,2,3,4,5
   c. 5
   d. 2
   e. 1,2,3,5

2. Endotoxins
   a. all have the same effect.
   b. are found in both gram-negative and positive
      bacteria.
   c. are part of the gram-positive cell wall only.
   d. are heat labile
   e. are proteins

3. Which of the following blood cells are the first to appear in response to an infection?
   a. eosinophils
   b. erythrocytes
   c. lymphocytes
   d. basophils
   e. neutrophils

4. Which of the following nonspecific substances is a normal component of serum?
   a. complement
   b. lysozyme
   c. interferon
   d. collagenase
   e. lysosome

5. Which of the following are ways in which normal flora protect against infection by pathogens?
   1. Normal flora provide a surface incompatible for attachment by the invader.
   2. Normal flora compete for essential nutrients, such as vitamins, amino acids, and iron.
   3. Normal flora produce metabolic by-products that are toxic.
   4. Normal flora produce lysozyme.
   5. There is no evidence that the normal flora protects against infection by pathogens.
      a. 2,3,4
      b. 1,2,3
      c. 1,2,3,4
      d. 4
      e. 5

6. A microorganism which is a commensal
   a. does not receive any benefit from its host.
   b. is harmful to its host.
   c. receives some benefit from its host.
   d. receives benefit at the same time the host benefits too.
   e. None of the above describe a commensal.

REVIEW QUESTIONS

1. Complete the following table by recording a + if the effect of the interaction is positive for either the microorganism or the human. Likewise, record a - if the interaction is negative (i.e., either one suffers from the association). Record a 0 if the interaction is neutral. Give an example of each type of interaction.

INTERACTION	MICROORGANISM	HUMAN	EXAMPLE
COMMENSALISM			
MUTUALISM			
PARASITISM			

2. Distinguish between infection and disease.

3. Explain the statement, "Normal floras are subject to change."

4. List Koch's postulates.

5. Compare and contrast four properties of endotoxins and exotoxins in the following table.

PROPERTY	ENDOTOXINS	EXOTOXINS
1.		
2.		
3.		
4.		

6. Describe how phagocytes destroy microorganisms.

NOTES

ANSWERS:
  Vocabulary: Terms and Definitions
    1. receptors 2. symbiosis 3. disease 4. normal flora 5. opportunists 6. virulence 7. infection
    8. transient flora 9. lipid A 10. ligands 11. pathogen
  Self-Test of Reading Material
    1. b 2. a 3. e 4. a 5. b 6. c

20 *EPIDEMIOLOGY AND PUBLIC HEALTH*

Epidemiology is the study of factors that influence the frequency and distribution of diseases. Exactly what does that mean! It means that epidemiologists are like detectives because they try to predict when an epidemic is occurring or is going to occur. They look for clues related to the number of new cases of a disease compared to its usual numbers. They look for clues as to where the epidemic started, where it will spread, and who are most likely to get the disease. From this information, public health services can try to interrupt the transmission of the disease by appropriate means. Epidemiologists need a lot of math and statistical training in addition to microbiology.

KEY CONCEPTS

1. Identifying the reservoir and mode of transmission of a disease agent can provide a means of preventing the disease.
2. Diseases in which symptomatic humans are the only reservoir are the easiest to control.
3. Worldwide travel and distribution of foods increases the global threat of disease.
4. Identifying the origin of an epidemic depends on a careful comparison of the characteristics and activities of those affected and unaffected by the epidemic.

SUMMARY

I. Epidemiology is the study of factors influencing the frequency and distribution of disease.
  A. Descriptive terms are important in the study of epidemiology.
    1. An epidemic is an unusually large number of cases in a population.
    2. Sporadic cases representing the usual incidence is known as endemic.
    3. A pandemic is a world wide epidemic.

II. Spread of Disease
  A. Humans, animals and nonliving environments can be reservoirs for disease agents.
  B. Diseases can be transmitted from person-to-person through direct contact, droplet transmission or indirect contact via inanimate objects.
  C. Pathogens can be transmitted in food or water.
  D. Pathogens can be transmitted through air.
  E. Arthropod vectors may transmit disease.
  F. The portal of entry may determine the outcome of disease.

III. Other Factors Play a Role in Disease Outcome and Spread
  A. The incubation period of a disease influences the extent of spread of the disease.
  B. The intensity of exposure to an infectious agent influences frequency, incubation period, and severity of disease.
  C. Immunization or prior exposure influences the susceptibility to infectious agents.
  D. Nutrition, crowding, and other environmental and cultural factors can shape the character of epidemics.
  E. The genetic background of an individual can influence susceptibility to infection.

IV. Epidemiological Studies
  A. Descriptive studies examine the persons, place and time of a disease outbreak.
  B. Analytical studies compare risk factors for developing disease in cases and controls.
  C. Precise identification of infectious agents is often necessary.

V. Infectious Disease Surveillance
   A. The public health network includes government agencies, hospital laboratories, and news media.
   B. The World Health Organization has targeted some diseases for eradication.

VI. Infectious Disease Control in Special Situations
   A. Special attention to good hygiene is essential in day care centers.
   B. Infectious diseases in hospitals

VOCABULARY: TERMS AND DEFINITIONS

_____ 1. An unusually large number of cases of a disease occurring in a given population

_____ 2. Infections contracted while a person is in the hospital

_____ 3. All of the places where an infectious agent is found

_____ 4. Scattered cases of disease occurring in a population over a long period

_____ 5. Inanimate objects that can serve to transmit infections

_____ 6. Viruses are used to identify bacterial strains

_____ 7. Food and other nonliving materials that may spread infections

_____ 8. The first person to have a disease in an epidemic

_____ 9. The study of the factors that influence the frequency and distribution of disease

_____ 10. An apparently healthy individual who harbors a pathogen

_____ 11. Antibiotic susceptibility patterns used to identify strains of bacteria

_____ 12. A worldwide epidemic

_____ 13. Insects or other animals that may spread disease

_____ 14. Using patterns of biochemical activity to identify bacterial strains

_____ 15. The number of cases developing per hundred people exposed

_____ 16. A chicken, for example, used to detect the presence of encephalitis-infected mosquitoes

a. epidemic
b. nosocomial infection
c. pandemic
d. carrier
e. endemic
f. fomites
g. epidemiology
h. attack rates
i. bacteriophage typing
j. index case
k. antibiograms
l. sentinel
m. biotyping
n. vehicles
o. reservoir
p. vectors

SELF-TEST OF READING MATERIAL

1. The study of when diseases occur and how they are transmitted is called
   a. a public health survey.
   b. epidemiology.
   c. ecology.
   d. the Centers for Disease Control.
   e. diagnostic microbiology.

2. Fomites are
   a. insect vectors.
   b. animate objects.
   c. an ancient tribe of Israel.
   d. inanimate objects.
   e. biological vectors.

3. A nosocomial infection is an infection
   a. that can come from the patient's own normal flora.
   b. acquired during hospitalization.
   c. always caused by medical personnel.
   d. a and b are correct.
   e. b and c are correct.

4. Which of the following are reservoirs for human infections.
   a. food and water
   b. humans
   c. animals
   d. All of the above are reservoirs for human infections.
   e. Only b and c are reservoirs.

5. A simple method for interrupting the person-to-person transmission of disease is
   a. immunization
   b. handwashing
   c. prophylactic chemotherapy
   d. isolation
   e. quarantine

6. Waterborne pathogens tend to be intestinal organisms.
   a. True
   b. False

7. Factors within a population that can influence the characteristic of an epidemic are
   1. age
   2. prior exposure
   3. crowding
   4. nutrition
   5. genetic background
   6. cultural factors

   a. 1,5
   b. 1,2,4,5
   c. 1,2,3,4,5
   d. All of the above.
   e. None of the above because only microbial factors can influence the characteristics of an epidemic.

8. A human carrier does not have an infection but simply spreads a disease through mechanical transmission.
   a. True
   b. False

9. People get diseases mostly from
   a. their pets.
   b. water.
   c. other people.
   d. food.
   e. air

10. Droplet nuclei are significant in the transmission of diseases of the
    a. digestive system.
    b. reproductive system.
    c. nervous system.
    d. respiratory system.
    e. skin.

REVIEW QUESTIONS

1. Distinguish among the terms vector, vehicle, carrier, and reservoir.

2. Describe the six components of body substance isolation.

3. Why are carriers such a public health problem?

4. Give two reasons why the attack rates for food-borne infections are usually much higher than those for waterborne infections.

5. Choose three characteristics of human populations that influence the nature of epidemics and give a brief explanation as to why.

ANSWERS:
Vocabulary: Terms and Definitions
   1. a 2. b 3. o 4. e 5. f 6. i 7. n 8. j 9. g 10. d 11. k 12. c 13. p 14. m 15. h 16. l
Self-Test of Reading Material
   1. b 2. d 3. d 4. d 5. b 6. a 7. d 8. b 9. c 10. d

NOTES

21 ANTIMICROBIAL AND ANTIVIRAL MEDICINES

Treatment of infectious disease has obviously been a primary goal of medicine for centuries. Paul Ehrlich's search for the "magic bullet" that would kill an infectious agent without harming the patient, was a systematic approach to the problem of antimicrobial therapy. The approach to the discovery of new antimicrobial agents today applies the knowledge of metabolism and genetics to solving the problem. Antimicrobial agents that interfere with processes unique to the pathogen will have the highest therapeutic index. While there are a considerable number of antimicrobial agents available, they fall into a few categories based either on their structure and/or their activity. This chapter summarizes the major groups of antimicrobial agents, their activity and application.

KEY CONCEPTS

1. Certain chemicals selectively kill or inhibit the growth of one organism while sparing another.
2. Many antimicrobial medicines exploit the differences that exist between prokaryotic and eukaryotic cell structure and metabolism.
3. Viruses and eukaryotic pathogens have fewer targets for selective toxicity.
4. Mutations and transfer of genetic information have allowed microorganisms to develop resistance to each new antimicrobial drug that has been developed.
5. Laboratory tests of the susceptibility of microorganisms to antimicrobial drugs help predict the effectiveness of the drugs in treating disease.

SUMMARY

I. Principle of Antimicrobial Chemotherapy
   A. Antimicrobial chemotherapy depends on the principle of selective toxicity.

II. Overview of Antibiotics
   A. The susceptibility of a specific antimicrobial depends on the clinical situation.
   B. Antimicrobials differ in their toxicity and spectrum of activity.
   C. Antimicrobials differ in how they are distributed, metabolized and excreted.
   D. Combinations of antimicrobial drugs can be antagonistic, synergistic, or additive.

III. Targets of Antimicrobial and Antiviral Drugs
   A. Differences between prokaryotic and eukaryotic cells are useful targets for antibacterial drugs.
      1. Cell wall synthesis
      2. Protein synthesis
      3. Nucleotide replication/transcription
      4. Cell membrane function
      5. Metabolic pathways
   B. Eukaryotic pathogens have many similarities to human cells so there are fewer targets for selective toxicity.
   C. Few targets exist for selective toxicity of antiviral drugs.

IV. Concerns in Using Antimicrobial Drugs
   A. Some antimicrobials may elicit allergic or other adverse reactions.
   B. Antimicrobials suppress the normal flora.

V. Bacterial Sensitivity Testing
   A. Susceptibility can be quantified by determining minimal inhibitory concentrations.
   B. The terms sensitive and resistant refer to whether a microorganism is likely to be treatable or not.

VI. Development of Resistance
    A. Some organisms are innately resistant to specific antibiotics.
    B. Previously sensitive bacteria can acquire antibiotic resistance.
    C. Acquisition of antimicrobial resistance can occur through spontaneous mutation.
    D. Acquisition of antimicrobial resistance can occur through DNA transfer.
    E. Control of antibiotic resistance relies on the cooperation of patients and physicians.

VII. Survey of Antimicrobial Medications
    A. Antibacterial drugs
    B. Antifungal drugs
    C. Antiviral drugs
    D. Drugs active against protozoan parasites

VOCABULARY: TERMS AND DEFINITIONS

The following list contains new terms introduced in this chapter. Use these terms to fill-in the blanks of the sentences that follow and you will have a definition or description of each new term.

antibiotics	R factors	resistance plasmids
pseudomembranous	secondary metabolites	minimal inhibitory
semisynthetic	therapeutic ratio	broad-spectrum
ß-lactamases	suicide	

1. Antibiotics in which a portion is made by a microorganism and the remainder is made by a chemist

    are called _____ antibiotics.

2. The lowest concentration of an antimicrobial agent capable of preventing growth is

    called the _____ concentration.

3. The _____ is defined as the highest dose a patient can tolerate without toxic effects, divided by the dose required to control the microbial infection.

4. Antibiotics are not directly involved in the growth of the organism that produces them and as such are known

    as _____ .

5. _____ antibiotics show activity against at least some strains of both gram-positive and gram-negative species of bacteria.

6. _____ colitis is an intestinal disease that is a potentially serious side effect of treatment with antibiotics.

7. _____ are enzymes that attack the ring structure of penicillin and its related antibiotics.

8. Plasmids that may contain genes for antimicrobial resistance to several antibiotics are

    called _____ or _____ .

9. _____ inhibitors bind to and irreversibly inactivate most ß-lactamases and are themselves inactivated in the process.

10. Chemical substances produced by some microorganisms that inhibit the growth of other microorganisms

are called _____ .

SELF-TEST OF READING MATERIAL

Use the list of antimicrobial agents on the right to match with the appropriate description on the left.

_____ 1. Inactivates ß-lactamases

_____ 2. Used to treat bladder infections caused by *E. coli*

_____ 3. Can cause pseudomembranous colitis

_____ 4. Major use as an alternative to penicillin

_____ 5. Used against both bacteria and protozoa

_____ 6. Toxic but lowers blood cholesterol

_____ 7. Useful only for influenza A

_____ 8. Attaches to sterols in eukaryotic cell membranes

_____ 9. Used for topical treatment and made by *Bacillus*

_____ 10. Can cause aplastic anemia

_____ 11. Cannot be taken with milk because it is inactivated by ions such as calcium

_____ 12. Used in combination with sulfas to treat urinary tract infections

_____ 13. Active against bacterial cell wall

_____ 14. Used to treat tuberculosis

_____ 15. One of only a few antibiotics used to treat *Pseudomonas aeruginosa*

a. sulfa drugs
b. trimethoprim
c. quinolones
d. isoniazid
e. metronidazole
f. penicillins
g. cephalosporins
h. clavulanic acid
i. tetracyclines
j. chloramphenicol
k. neomycin
l. erythromycin
m. lincomycin
n. bacitracin
o. amphotericin B
p. amantadine

REVIEW QUESTIONS

1. What three groups of microorganisms produce antibiotics? Give an example of an organism from each group and an antibiotic it produces.

2. Outline the steps of the production of a new antibiotic.

3. What is clavulanic acid and what value is it in treating bacterial infections?

4. Give an example of an antifungal, an antiprotozoan, and an antiviral medicine. Why are infections by fungi and protozoa often more difficult to treat than bacterial infections?

5. What is the basis of selective toxicity and how does it relate to the therapeutic index?

ANSWERS:

Vocabulary: Terms and Definitions

1. semisynthetic 2. minimal inhibitory 3. therapeutic ratio 4. secondary metabolites 5. broad-spectrum
6. pseudomembranous 7. ß-lactamases 8. R/resistance 9. suicide 10. antibiotics

Self-Test of Reading Material

1. h 2. a 3. m 4. l 5. e 6. k 7. p 8. o 9. n 10. j 11. i 12. b 13. f,g 14. d 15. c

NOTES

NOTES

22 *SKIN INFECTIONS*

We usually do not think of it in that way, but our skin is our largest organ. It is the boundary between our body and the outside world. We all learned in sixth grade hygiene class that our skin was our first line of defense against infection. This chapter describes a few of the things that can invade and colonize our body when the skin barrier is broken. In a few instances the skin is attacked from the inside by organisms entering through the respiratory or digestive systems. You will be introduced to your skin and its normal flora so that you will have a better appreciation of the truly remarkable job it does for you every day.

KEY CONCEPTS

1. The skin is a large complex organ covering the external body surfaces.
2. Properties of the skin cause it to resist microbial colonization, provide a physical barrier to the entry of potential pathogens, and assist the body's regulation of temperature and fluid balance.
3. The changes in the skin during an infectious disease commonly reflect disease involvement of other parts of the body.

SUMMARY

I. Anatomy and Physiology
   A. The skin repels potential pathogens by shedding and being dry, acidic and toxic.

II. The Normal Flora Of Human Skin Consistently Includes Three Groups Of Microorganisms
   A. Diphtheroids are gram-positive, pleomorphic, rod-shaped bacteria.
      1. Growth of *Propionibacterium acnes* is enhanced by oily secretion of sebaceous glands. *P. acnes* produces fatty acids from the gland secretion. They are usually anaerobes, but some are able to tolerate growth in air.
   B. Staphylococci are gram-positive cocci that help prevent colonization by potential pathogens and maintain the balance among flora of the skin.
      1. *Staphylococcus epidermidis* is universally present on the skin and is pathogenic at times.
   C. *Malassezia* are single-celled yeasts found universally on the skin, probably cause some cases of dandruff as well as tinea versicolor, and more serious rashes in AIDS patients.

III. Skin Diseases Caused By Bacteria
   A. Furuncles and carbuncles are caused by *Staphylococcus aureus,* which is coagulase positive and is often resistant to penicillin and other antibiotics. Carbuncles are dangerous because the infection may be carried to heart, brain or bones.
   B. Staphylococcal scalded skin syndrome results from exotoxins produced by certain strains of *Staphylococcus aureus.*
   C. Impetigo is a superficial skin infection caused by *Staphylococcus aureus* and *Streptococcus pyogenes.*
      1. The capsule and the M-protein of the cell wall of S. *pyogenes* interfere with phagocytosis, thus aiding pathogenesis.
      2. Penicillin or erythromycin is effective in most S. *pyogenes* infections.
      3. Acute glomerulonephritis, a kidney disease, is a complication of some S. *pyogenes* infections.
   D. Rocky Mountain spotted fever (RMSF), caused by the obligate intracellular bacterium, *Rickettsia rickettsii,* is transmitted to humans by the bite of an infected tick. Tetracycline or chloramphenicol is used in the treatment of RMSF.
   E. Lyme disease can imitate many other diseases. It is caused by a spirochete, *Borrelia burgdorferi* and is transmitted to humans by ticks. Erythema chronicum migrans is the hallmark of the disease. Doxycycline, amoxicillin, and erythromycin are effective in treating early stages of the disease.

IV. Skin Diseases Caused By Viruses
  A. Chickenpox is a common disease of childhood caused by the varicella-zoster virus, a member of the herpesvirus family.
    1. Herpes zoster or "shingles" can occur months or years after chickenpox and represents a reactivation of a varicella virus infection latent in a sensory nerve.
    2. Herpes zoster cases can be responsible for chickenpox epidemics.
  B. Rubeola (measles) is a potentially dangerous viral disease that can be fatal in some cases and can lead to serious secondary bacterial infections in others.
    1. Rubeola can be controlled by immunizing young children and susceptible adults with a live attenuated vaccine. A two-dose regimen is now employed in an effort to decrease the size of the nonimmune population.
  C. Rubella (three-day measles, German measles) contracted by a woman in the first eight weeks of pregnancy results in at least a 90% chance of congenital rubella syndrome. Immunization with a live virus protects against this disease, but many women still reach childbearing age without proper vaccination.
  D. Numerous viruses can cause rashes. In recent years the viral cause of two exanthems has been determined.
    1. Erythema infectiosum is characterized by a "slapped cheek" rash. It is caused by parvovirus B-19.
    2. Exanthem subitum is characterized by several days of high fever and a transitory rash appearing as the temperature returns to normal. The disease is caused by human herpesvirus type 6.
  E. Warts are skin tumors caused by a number of papillomaviruses. They are generally benign, but some genital warts have been associated with cancer of the cervix.

V. Skin Diseases Caused by Fungi
  A. Invasive skin infections are sometimes caused by the yeast, *Candida albicans*.
  B. Athlete's foot and ringworm are caused by certain species of mold-type fungi which feed on keratin-containing cells.

VOCABULARY: TERMS AND DEFINITIONS

The following list contains new terms introduced in this chapter. Use these terms to fill-in the blanks of the sentences that follow and you will have a definition or description of each new term.

zoonosis	M-protein	coagulase
protein	epidermis	exanthans
pyoderma	penicillinase	dermatophytes
sebum	*Staphylococcus aureus*	

1. _____ is an oily secretion produced glands in the skin that empty into the sides of the hair follicles.

2. Skin rashes are also called _____ .

3. _____ is a cell wall component of virtually all strains of *Staphylococcus aureus* and binds to the Fc portion of antibody molecules.

4. A skin infection characterized by superficial pus-filled blister is called _____ .

5. _____ is the surface layer of the skin which is composed of scaly material.

6. A penicillin-destroying enzyme produced by many strains of *Staphylococcus aureus* is _____ .

7. A _____ is a disease that exists primarily in animals other than humans but that can be transmitted to humans.

8. Skin-invading molds are collectively called _____ .

9. A cell wall component of *Streptococcus pyogenes* that inhibits phagocytosis is _____ .

10. _____ is an enzyme that causes plasma to coagulate and is used to identify pathogenic

strains of _____ .

SELF-TEST OF READING MATERIAL

Match the condition on the left with its etiological agent on the right. A condition may have more than one causative agent and an agent may cause more than one condition.

_____ 1. furuncle

_____ 2. warts

_____ 3. scalded skin syndrome

_____ 4. fifth disease

_____ 5. chicken pox

_____ 6. food poisoning

_____ 7. subacute sclerosing panencephalitis

_____ 8. impetigo

_____ 9. measles

_____ 10. scarlet fever

_____ 11. shingles

_____ 12. toxic shock syndrome

_____ 13. Lyme disease

_____ 14. glomerulonephritis

_____ 15. Rocky mountain spotted fever

a. *Staphylococcus aureus*
b. papillomavirus
c. B-19
d. *Streptococcus pyogenes*
e. Rubeola
f. *Rickettsia rickettsii*
g. Herpes zoster
h. *Borrelia burgdoferi*
i. Varicella zoster

16. The layer of skin that contains many nerves, blood vessels and lymphatic vessels is the
    a. epidermis
    b. dermis
    c. hypodermis
    d. keratin
    e. scaly layer

17. Vaccination of infants against rubeola is done at about 15 months of age because of
    a. placentally transferred maternal antigens.
    b. infant antibodies.
    c. it is easier to give the injection.
    d. infant antigens.
    e. placentally transferred maternal antibodies.

18. Which of the following allows *Staphylococcus aureus* to colonize oily hair follicles?
    a. lipase
    b. protein A
    c. collagenase
    d. coagulase
    e. hyaluronidase

19. An epidermolytic toxin is involved in
    a. chicken pox
    b. shingles
    c. scalded skin syndrome
    d. exanthem subitum
    e. warts

20. Lyme disease is a tick-borne disease with similarities to syphilis.
    a. True
    b. False

REVIEW QUESTIONS

1. Describe, in addition to protection against infection, six other functions of the skin.

2. Give four characteristics of skin that help it resist infection.

3. How does the normal flora of the skin help protect against infection?

4. Describe four extracellular products of *Staphylococcus aureus* and how they contribute to its virulence.

5. Rocky Mountain spotted fever and Lyme disease share some common features. Describe two they have in common.

6. What is the relationship between varicella and herpes zoster?

7. How are warts transmitted?

ANSWERS:
  Vocabulary: Terms and Definitions
    1. sebum 2. exanthems 3. protein A 4. pyoderma 5. epidermis 6. penicillinase 7. zoonosis 8. dermatophytes
    9. M protein 10. coagulase/*Staphylococcus aureus*
  Self-Test of Reading Material
    1. a 2. b 3. a 4. c 5. i 6. a 7. e 8. a,d 9. e 10. d 11. g 12. a 13. h 14. d 15. f 16. b 17. e 18. a 19. c
    20. a

NOTES

23 UPPER RESPIRATORY SYSTEM INFECTIONS

We all know that treating a cold will cause it to go away in seven days but if you leave it alone it will disappear in a week. Like colds, most upper respiratory infections are not life-threatening, but can be extremely uncomfortable and irritating. We take into our lungs very large volumes of air each day and some of that air will contain infectious agents. Our upper respiratory tract is a marvel at preventing infections, most of the time. When we compromise the barriers that our system provides with abuse such as smoking and narcotics, we make ourselves more susceptible to infection. Serious infections such as diphtheria can be controlled by immunization but strep throat must be treated to prevent serious sequelae.

KEY CONCEPTS

1. Untreated skin or throat infections caused by *Streptococcus pyogenes* can occasionally result in serious kidney or heart damage even though these organs are not infected. The damage is probably caused by the immune system.
2. Some diseases arise because an exotoxin is absorbed from a localized infection. The toxin then circulates throughout the body causing damage to tissues some distance from the site of infection.
3. Circulating toxins may not affect all tissues equally. A toxin may target specific organs because it attaches to receptors on the cells of the organ.
4. Many different kinds of infectious agents can produce the same symptoms and signs of disease.

SUMMARY

I. Anatomy and Physiology
   A. The moist linings of the eyes (conjuctiva), the nasolacrimal duct, the middle ears, sinuses, mastoid air cells, nose and throat comprise the main structures of the upper respiratory system.
   B. An important defense system is the movement of mucus by the ciliated cells that line much of the respiratory system. The middle ears, mastoids, and sinuses are usually kept free of microorganisms by this mechanism.
   C. The functions of the upper respiratory tract include temperature and humidity regulation of inspired air, and removal of microorganisms.

II. Normal Flora
   A. Secretions of the nasal entrance are often colonized by diphtheroids, micrococci, and *Staphylococcus aureus* (coagulase-positive staphylococci).
   B. The nasopharyngeal flora includes alpha-hemolytic streptococci, nonhemolytic streptococci, *Moraxella catarrhalis,* diphtheroids and anaerobes of the genus *Bacteroides.*

III. Bacterial Infections
   A. *Streptococcus pyogenes* causes strep throat, a very important bacterial infection of the throat. Infection may lead to scarlet fever, rheumatic fever, toxic shock, or glomerulonephritis.
   B. Diphtheria, caused by *Corynebacterium diphtheriae,* is a toxin-mediated disease that can be prevented by immunization.
   C. Otitis media and sinusitis develop when infection extends from the nasopharynx.
   D. Conjunctivitis (pink eye) is usually caused by *Haemophilus influenzae* or *Streptococcus pneumoniae* . Viral causes, including adenoviruses and rhinoviruses, usually result in a milder illness.

IV. Viral Infections
   A. The common cold can be caused by many different viruses, rhinoviruses being the most common.
   B. Adenoviruses cause illnesses varying from mild to severe. They may resemble a common cold or strep throat.

SELF-TEST OF READING MATERIAL

1. Which of the following is a serious complication that may develop after recovery from strep throat?
   a. scarlet fever
   b. acute glomerulonephritis
   c. rheumatic fever
   d. a and b are correct.
   e. b and c are correct.

2. The rash of scarlet fever is caused by
   a. an erythrogenic toxin.
   b. bacteria in the skin.
   c. high fever.
   d. a virus.
   e. an allergy.

3. A toxoid is a(n)
   a. antibody against a toxin.
   b. inactivated toxin.
   c. larger molecular weight toxin.
   d. small molecular weight toxin.
   e. Two of the above are correct.

4. Tonsils are lymphoid tissue and important producers of immunity to infectious agents.
   a. True
   b. False

5. Which of the following will interfere with normal ciliary action of the respiratory epithelium and make an individual more susceptible to infection?
   a. viral infection
   b. alcohol
   c. narcotics
   d. tobacco
   e. All of the above promote infection.

6. Most strains of *Streptococcus pyogenes* produce ß-hemolysis on blood agar plates from the activity of two hemolytic enzymes that are sensitive to oxygen.
   a. True
   b. False

7. New tests for strep throat that are rapid and can be performed in a physician's office rely on *Streptococcus pyogenes*
   a. antibodies in the throat.
   b. antigens in the blood.
   c. cultivation.
   d. antibodies in the blood.
   e. antigens in the throat.

8. Acute glomerulonephritis, a consequence of strep throat infections, is actually caused by
   a. immune complexes in the kidneys.
   b. *Streptococcus pyogenes* endotoxins.
   c. immune complexes in the throat.
   d. *Streptococcus pyogenes* exotoxins.
   e. a and d are correct.

9. Which of the following are most seriously affected by rheumatic fever.
   a. heart
   b. kidney
   c. brain
   d. liver
   e. throat

10. Depletion of which of the following enhances toxin synthesis in *Corynebacterium diphtheriae*?
   a. oxygen
   b. serum
   c. iron
   d. vitamins
   e. carbon dioxide

11. *Streptococcus pneumoniae* and *Haemophilus influenzae* are the most important bacterial pathogens associated with otitis media or middle ear infection.
    a. True
    b. False

12. *Streptococcus pneumoniae* and *Haemophilus influenzae* are common causes of bacterial conjuctivitis or pink eye.
    a. True
    b. False

REVIEW QUESTIONS

1. List the genera of normal flora of the upper respiratory tract. Identify and name those that contain pathogenic species.

2. Explain the designation of *Streptococcus pyogenes* as the "group A ß-hemolytic strep."

3. *Streptococcus pyogenes* cannot be isolated from patients with acute glomerulonephritis or rheumatic fever but it is still considered the "cause" of these conditions. Explain this observation.

4. Give two examples of diseases whose symptoms are caused by exotoxins produced by lysogenized bacterial cells. Also, write the correct scientific name of each of the causative agents.

5. What are the functions of the A and the B fragment of the diphtheria toxin?

6. Why is a vaccine against *Streptococcus pyogenes* unlikely?

ANSWERS:
  Self-Test of Reading Material
     1. e 2. a 3. b 4. a 5. e 6. b 7. e 8. a 9. a 10. c 11. a 12. a

---

24 *LOWER RESPIRATORY SYSTEM INFECTIONS*

The lower respiratory tract is usually considered free from normal flora and usually only contains relatively few transient organisms that have been inhaled. Infection of the lower respiratory tract is known generally as pneumonia and can be caused by bacterial, viral and fungal agents. Other irritations, such as those from chemicals, are also referred to as pneumonias. Many diseases, such as measles and chicken pox, first enter our body through the respiratory system and are often spread to others by respiratory droplets. In addition to several bacterial pneumonias, other serious diseases of the lower respiratory tract include tuberculosis, Legionnaires' disease, and pertussis.

KEY CONCEPTS

1. Very efficient mechanisms protect the lower respiratory system; bacterial lung infections usually occur only in people whose defenses are impaired.
2. Most pneumonias are bacterial or viral, but eukaryotic microorganisms, chemicals, and allergies can also cause pneumonias.
3. A vaccine composed of killed organisms can confer protection against some gram-negative bacterial diseases such as pertussis.
4. Most *Mycobacterium tuberculosis* infections become latent, posing a risk of reactivation throughout life.
5. Most deaths from influenza are caused by secondary bacterial infections.
6. The host's immune system can play a key role in the pathogenesis of lung disease.

SUMMARY

I. Anatomy and Physiology
    A. The lower respiratory system includes the trachea, bronchi, bronchioles, and alveoli. Pleural membranes cover the lungs and line the chest cavity.
    B. Ciliated cells line much of the respiratory tract and remove microorganisms by constantly propelling mucus out of the respiratory system.

II. Bacterial Pneumonias
    A. Bacterial pneumonias can be caused by gram-positive and gram-negative organisms and one species of *Mycoplasma.*
        1. *Streptococcus pneumoniae* is one of the most common causes of pneumonia. The bacterium is also known as the pneumococcus.
        2. *Klebsiella pneumoniae* may cause permanent damage to the lung. Serious complications are more common than with other bacterial pneumonias. Treatment is more difficult, partly because klebsiellas often contain R-factor plasmids.
        3. *Mycoplasma* pneumonia is often mild and referred to as "walking pneumonia." Serious complications are rare. Penicillins and cephalosporins are not useful in treatment because *M. pneumoniae* lacks a cell wall. Cold agglutinins often help confirm the diagnosis.

III. Other Bacterial Infections of the Lung
    A. Pertussis (whooping cough) is characterized by violent spasms of coughing and gasping. The cause is a gram-negative rod, *Bordetella pertussis.* Childhood immunization prevents the disease.
    B. Tuberculosis is generally slowly progressive, or heals and remains latent, presenting the risk of reactivation. The cause is an acid-fast rod, *Mycobacterium tuberculosis.*
    C. Legionnaires' disease occurs when there is a high infecting dose of microorganisms, or underlying lung disease. The cause, *Legionella pneumophila,* a rod-shaped bacterium, is commonplace in the environment. It requires special stains and culture techniques for diagnosis.

IV. Viral Infections of the Lower Respiratory Tract Caused by Orthomyxoviruses.
   A. Serious epidemics are characteristic of influenza A viruses. Antigenic shifts and drifts are responsible.
   B. Deaths are usually but not always caused by secondary infection.
   C. Antibody against hemagglutinin gives protection.
   D. Reye's syndrome may rarely occur during recovery from influenza B and other viral infections, but is probably not caused by the virus itself.

V. Fungal Infections of the Lung
   A. Coccidioidomycosis occurs in hot, dry areas of the Western Hemisphere. The airborne spores of the dimorphic soil fungus, *Coccidioides immitis,* cause the infection.
   B. Histoplasmosis is similar to coccidioidomycosis but occurs in tropical and temperate zones around the world. The causative fungus, *Histoplasma capsulatum,* is dimorphic and found in soils contaminated by bat or bird droppings.

VOCABULARY: TERMS AND DEFINITIONS

The following list contains new terms introduced in this chapter. Use these terms to fill-in the blanks of the sentences that follow and you will have a definition or description of each new term.

neuraminidase	aerosols	tuberculin
spelunker's	tubercles	sputum
mycoses	acellular	Mantoux
caseous necrosis	antigenic shift	hemagglutinin
antigenic drift		

1. Tiny liquid or solid particles suspended in air and are often highly contagious are called _____ .

2. _____ is a thick discharge that accumulates during respiratory disease and consists of mucus and a variety of cells.

3. A purified protein fraction isolated from *Mycobacterium tuberculosis* and used to test for cases of tuberculosis is called _____ and the test is known as the _____ test.

4. _____ are infections caused by fungi.

5. The glycoprotein spikes on the outer membranes of influenza viruses are called _____ and _____ .

6. Vaccines which contain the antigens that stimulate immunity but lack other components of the bacterial cell are called _____ vaccines.

7. Histoplasmosis has been called the _____ disease because cave explorers are at a risk for contracting the condition.

8. _____ and _____ are types of genetic variation that are responsible for changes in the antigens of influenza viruses.

9. The granulomas of tuberculosis are called _____ which in some cases cause the death of the tissue and form a cheesy material by a process called _____ .

SELF-TEST OF READING MATERIAL

1. Walking pneumonia is caused by
   a. *Mycoplasma tuberculosis.*
   b. *Mycobacterium pneumoniae.*
   c. *Klebsiella pneumoniae.*
   d. *Mycobacterium tuberculosis.*
   e. *Mycoplasma pneumoniae.*

2. Histoplasmosis is associated with the growth of *Histoplasma capsulatum* in
   a. soil containing bird and bat droppings.
   b. dust and dry air.
   c. plants.
   d. hospitals.
   e. nasal secretions.

3. Deaths from influenza are most often due to
   a. heart failure.
   b. secondary bacterial infections.
   c. exotoxins.
   d. kidney failure
   e. suffocation.

4. Which of the following organisms grow naturally inside amoebas as well as body phagocytes?
   a. *Mycobacterium tuberculosis*
   b. *Mycoplasma pneumoniae*
   c. influenza A
   d. *Legionella pneumophila*
   e. *Klebsiella pneumoniae*

5. *Mycoplasma pneumoniae* infections are most common in
   a. people over the age of 40.
   b. infants.
   c. adolescents.
   d. people over the age of 60.
   e. people between 40 and 60.

6. Infection with *Bordetella pertussis* is confined to lobes of the lungs.
   a. True
   b. False

7. Person-to-person transmission is not characteristic of infections caused by
   a. *Legionella pneumophila*
   b. *Mycoplasma pneumoniae*
   c. influenza A
   d. *Klebsiella pneumoniae*
   e. *Mycobacterium tuberculosis*

8. Reye's syndrome sometimes occurs with infections of influenza B.
   a. True
   b. False

9. The "P" in DPT is
   a. attenuated *Bordetella pertussis*
   b. pertussis exotoxin.
   c. killed *Bordetella pertussis.*
   d. pertussis endotoxin.
   e. toxoid.

10. *Mycobacterium tuberculosis* endospores are readily killed by pasteurization.
    a. True
    b. False

REVIEW QUESTIONS

1. Why is the BCG vaccine not used in AIDS patients?

2. Why is "walking" pneumonia not treated with penicillin?

3. What accounts for the "swine" in swine flu?

4. Why did it takes several months to discover the agent that was responsible for the outbreak of Legionnaires' disease in 1976?

5. What factors are contributing to the increase of tuberculosis?

6. How do alcoholism and cigarette smoking predispose a person to pneumonia?

ANSWERS:

Vocabulary: Terms and Definitions

1. aerosols 2. sputum 3. tuberculin/Mantoux 4. mycoses 5. hemagglutinin/neuraminidase 6. acellular
7. spelunker 8. antigenic drift/antigenic shift 9. tubercles/caseous necrosis

Self-Test of Reading Material

1. e 2. a 3. b 4. d 5. c 6. b 7. a 8. a 9. c 10. b

NOTES

25 *UPPER ALIMENTARY SYSTEM INFECTIONS*

The upper alimentary tract includes the organs from the mouth down to the stomach along with accessory structures such as the salivary glands. Much of the alimentary tract itself has an associated normal flora and represents one of the major routes of entry into the body by pathogens and opportunists. Our food and water contain a large variety of organisms that find their way into the body by this route. The stomach with its acid contents is an extremely efficient barrier to many microorganisms reaching the lower alimentary tract. The stomach was considered not too many years ago as being essentially sterile and not a usual site of infection. Infection of the stomach is now known to be commonplace.

KEY CONCEPTS

1. The alimentary tract is a major route for the entry of pathogenic microorganisms into the body.
2. The pathogenesis of tooth decay depends on both diet and acid-forming bacteria, which are adapted to colonize hard, smooth surfaces within the mouth.
3. Infections of the esophagus are so unusual in normal people that their occurrence suggests immunodeficiency.
4. A properly functioning stomach destroys most microorganisms before they reach the intestine.

SUMMARY

I. Anatomy and Physiology
   A. The upper alimentary system is composed of the mouth and salivary glands, esophagus, and stomach.
   B. Degradation of complex foods begins in the mouth with the grinding action of teeth and mixing with saliva. Saliva is produced by glands in the floor of the mouth and on the sides of the face. It contains the enzyme amylase that begins the degradation of starch. It protects the teeth by several mechanisms.
   C. Three different kinds of cells line the stomach. One produces hydrochloric acid; another, enzymes; and the third, mucus. One of the important stomach enzymes is pepsin, which begins the degradation of proteins under acid conditions.

II. Normal Flora
   A. The species of bacteria that inhabit the mouth colonize different locations depending on their ability to attach to structures such as the tongue, cheeks, teeth, or other bacteria.
   B. Dental plaque consists of enormous quantities of bacteria of various species attached to teeth or to each other. Streptococci and actinomycetes are generally responsible for the initiation of plaque formation.
   C. The presence of teeth allows for colonization by anaerobic bacteria.

III. Infections Caused by Oral Bacteria
   A. Dental caries is an infectious disease caused mainly by *Streptococcus mutans* and related species.
   B. Pathogenesis of cariogenic dental plaque involves formation by *S. mutans* of extracellular glucans from dietary sucrose.
   C. Penetration of the calcium phosphate tooth structure depends on acid production by cariogenic dental plaque. Prolonged acid production is fostered by the ability of *S. mutans* to form intracellular polysaccharide.
   D. Control of dental caries depends mainly on supplying adequate fluoride and restricting dietary sucrose. Dental sealants are important in children.
   E. Periodontal disease is mainly responsible for tooth loss in older people.
   F. Endocarditis in patients with abnormal heart valves is commonly due to oral bacteria. Entry of these bacteria

into the blood stream frequently occurs with dental procedures.

IV.   Viral Diseases of the Mouth and Salivary Glands

    A.   Herpes simplex is caused by an enveloped DNA virus that can be readily cultivated in tissue cell cultures. Infected cells show intranuclear inclusion bodies.

    B.   Latent infections are characteristic of herpesviruses. HSV persists inside sensory nerves in a noninfectious form that can become infectious and produce active disease when the body undergoes certain stresses.

    C.   Mumps is caused by an enveloped RNA virus that is prone to infect the parotid glands, meninges, testicles, and other body tissues.

    D.   Mumps virus generally causes more severe disease in persons beyond the age of puberty.

    E.   Mumps can be prevented using a live vaccine.

V.   Infections of the Esophagus and Stomach

    A.   Esophageal infections are seen mainly in immunocompromised individuals. The main causes are herpes simplex virus and the yeast, *Candida albicans*.

    B.   *Helicobacter pylori* gastritis is a common infection of the stomach.

        1.   *H. pylori* predisposes the stomach and upper duodenum to peptic ulcers.

        2.   Treatment is difficult even though the organism is susceptible to antimicrobial medications *in vitro*. However, effective treatment prevents peptic ulcer recurrence.

VOCABULARY: TERMS AND DEFINITIONS

The following list contains new terms introduced in this chapter. Use these terms to fill-in the blanks of the sentences that follow and you will have a definition or description of each new term.

lactoferrin	**endoscopy**	**pylorus**
herpetic whitlow	**gingivitis**	**calculus**
oxidation-reduction	**glucan**	**dental plaque**
periodontal disease		

1. The name for calcified plaque on the teeth is _____ .

2. _____ is a substance in saliva that binds to iron and removes it from use by bacteria.

3. A low _____ potential means that little or no oxygen is available for growth of bacteria.

4. _____ is the inspection of an internal area of the body with the use of a flexible fiber-optic instrument.

5. Insoluble polysaccharides composed of repeating subunits of glucose are called _____ .

6. The _____ is a muscular valve which determines the rate at which the stomach contents enter the intestine.

7. Masses of bacteria attached to the teeth by various polymeric substances are called _____ .

8. _____ is a painful finger infection caused by Herpes simplex virus and an occupational hazard among dentists.

9. _____ is a chronic inflammatory process involving the tissues around the roots of the teeth.

10. An inflammation of the gums is called _____ .

SELF-TEST OF READING MATERIAL

1. The teeth are made up largely of
   a. urease.
   b. hydroxyapatite crystals.
   c. enamel.
   d. dentin.
   e. amylase.

2. Herpes simplex virus can remain latent in a noninfectious form within
   a. the lips.
   b. the mouth.
   c. saliva.
   d. sensory nerves.
   e. plaque.

3. *Helicobacter pylori* infection can be diagnosed from the presence of which of the following in stomach biopsy?
   a. amylase
   b. intranuclear inclusions
   c. urease
   d. hydroxyapatite
   e. acid

4. Which of the following is commonly associated with recurrences of herpes simplex virus?
   a. fever
   b. sunburn
   c. menstruation
   d. All of the above are associated with recurrences.
   e. Only a and b are associated with recurrences.

5. Sterility is common in older men in which the mumps have "dropped" and infected the testicles.
   a. True
   b. False

6. Which of the following is associated with an increased incidence of stomach cancer?
   a. herpes simplex virus
   b. mumps
   c. halitosis
   d. *Helicobacter pylori*
   e. herpetic whitlow

7. Ancyclovir does not effect latent herpes simplex virus so it does not cure the disease but it can relieve the symptoms and decrease the amount of virus being shed.
   a. True
   b. False

8. The severe swelling and pain associated with mumps is due to
   a. endotoxins.
   b. inflammatory response.
   c. immune complexes.
   d. exotoxins.
   e. Two of the above are correct.

9. Mumps initially reproduces in the
   a. blood system.
   b. nervous system.
   c. salivary glands.
   d. digestive system.
   e. respiratory system.

10. Mumps vaccine is recommended for those infected with HIV because the benefit far outweighs the risk from the vaccine.
    a. True
    b. False

REVIEW QUESTIONS

1. Why is penicillin started before a dental procedure on someone that has known abnormal heart valves?

2. Since only one serotype of mumps virus is known that makes it a good candidate for eradication. Explain.

3. Why must teeth be present in order for anaerobes to colonize the mouth?

4. Describe how mumps virus results in swelling of the salivary glands?

ANSWERS:
Vocabulary: Terms and Definitions
1. calculus 2. lactoferrin 3. oxidation-reduction 4. endoscopy 5. glucan 6. pylorus 7. dental plaque
8. herpetic whitlow 9. periodontal disease 10. gingivitis
Self-Test of Reading Material
1. b 2. d 3. c 4. d 5. b 6. d 7. a 8. b 9. e 10. a

26 LOWER ALIMENTARY SYSTEM INFECTIONS

Gastrointestinal infections of the lower alimentary tract are among the most common and often most severe infections of humans. The stomach protects against a large number of pathogens, but many have adapted means to get through the stomach unharmed. Many parasites, such as worms and protozoa, even use the acid condition of the stomach to initiate their exit from protective eggs and cysts. Diseases of the lower alimentary tract are usually transmitted by the fecal-oral route meaning that the next host gets the disease from eating food or drinking water, or putting contaminated items such as fingers into their mouth. The lower bowl is home to a variety of bacteria belonging to the normal flora. Diarrhea, from simple to invasive forms, is responsible for millions of deaths worldwide each year.

KEY CONCEPTS

1. Some microbial toxins alter the secretory function of cells in the small intestine without killing or visibly damaging them.
2. Unsuspected human carriers can remain sources of enteric infections for many years.
3. Viruses, bacteria, and protozoa are important causes of human intestinal diseases.
4. Food poisoning can be caused either by infectious microorganisms consumed with the food or toxic products of microbial growth in food.

SUMMARY

I. Anatomy and Physiology of the Lower Alimentary System
   A. Important functions include breakdown of food macromolecules to their subunits, absorbing nutrients, and recycling fluids.
   B. Bile, produced by the liver, is inhibitory for many bacteria, and aids digestion of fats and vitamins. It gives the brown color of feces when acted on by intestinal bacteria.
   C. The pancreas produces insulin and alkaline fluid containing digestive enzymes.
   D. The small intestine secretes digestive juices and absorbs nutrients. It has a large surface area.
   E. The large intestine contains large numbers of microorganisms, many of which are opportunistic pathogens. It helps recycle body water by reabsorption of water from the feces.

II. Normal Flora of the Digestive Tract
   A. The small intestine has few microorganisms.
   B. Microorganisms make up about one-third of the weight of feces found in the large intestine.
   C. Anaerobes of the genera *Lactobacillus* and *Bacteroides* are the most prevalent.
   D. The biochemical activities of microorganisms in the digestive tract include synthesis of vitamins, degradation of indigestible substances, competitive inhibition of pathogens, production of cholesterol, chemical alteration of medications, and carcinogen production.

III. Bacterial Disease of The Digestive Tract
   A. Diarrhea is a major cause of death worldwide.
   B. Cholera is a severe form of diarrhea caused by a toxin of *Vibrio cholerae* that acts on the epithelium of the small intestine.
   C. Species of *Shigella* are common causes of dysentery because they invade the colon epithelium.
   D. Gastroenteritis is often caused by salmonellas of animal origin. The organisms often enter the intestinal tract with food, usually eggs and poultry.
   E. Gastroenteritis can be caused by certain strains of *Escherichia coli*. However, only strains possessing virulence factors cause disease. These factors often depend on plasmids. Plasmids and therefore virulence, can be transferred to other enteric organisms.

F.   *Campylobacter jejuni* is a common bacterial cause of diarrhea in the United States. Like the salmonellas, it usually originates from domestic animals. Like the shigellas and certain strains of *E. coli,* it can cause dysentery.

G.   Typhoid fever is caused by *Salmonella typhi*
1.   The disease is characterized by high fever, headache, and abdominal pain. Untreated, it has a high mortality rate.
2.   Human carriers maintain the disease.
3.   A new oral attenuated vaccine helps in prevention.

H.   Food Poisoning
1.   Illnesses are caused by toxic products of microbial growth in food.
2.   Staphylococcal food poisoning, caused by certain strains of *Staphylococcus aureus,* is one of the most common forms. The toxin is heat-stable.
3.   Food poisoning is also caused by the spore-forming gram-positive rods, *Clostridium perfringens* and *Bacillus cereus.*
4.   Botulism is characterized by paralysis and occurs in three forms: food-borne, intestinal, and wound.

IV.   Poisoning caused by fungi
A.   Growth of certain species of fungi on grain can cause ergot poisoning.
B.   Aflatoxins are fungal poisons produced by molds of the genus *Aspergillus.* Liver damage and certain cancers can result from ingestion of aflatoxins.

V.   Viral Infections of The Lower Alimentary System
A.   Viral Hepatitis
1.   Hepatitis A virus (HAV) is spread by fecal contamination of hands, food, or water. Most cases are mild and may even be asymptomatic. Antibody to the virus is widespread in the population.
2.   Hepatitis B virus (HBV) is spread mainly by blood and blood products but also by sexual intercourse, and from mother to infant. HBV is generally more severe than HAV. Chronic infection is fairly frequent and can lead to scarring of the liver and liver cancer. Carriers are common and may have infectious virus in their bloodstreams for years without knowing it.
3.   Other hepatitis viruses:
a.   Hepatitis C virus (HCV) causes most of the cases of transfusion-associated hepatitis.
b.   Hepatitis delta virus (HDV) is a defective RNA virus that requires the presence of HBV to replicate.
c.   Hepatitis E virus (HEV), an RNA virus, is transmitted by the fecal-oral route and can cause fatalities, especially in pregnant women.
d.   Hepatitis G virus (HGV) causes some cases of transfusion-associated hepatitis.
B.   Viral Gastroenteritis
1.   Rotoviruses and Norwalk viruses are the most important causes of this disease. Rotoviruses affect mainly children, but also can cause traveler's diarrhea. Norwalk viruses cause about half of the gastroenteritis outbreaks in the United States.

VI.   Protozoan Diseases of the Lower Alimentary System
A.   Giardiasis is caused by *Giardia lamblia* which is usually transmitted by drinking water contaminated by feces of humans and wild animals. It is distributed worldwide and is a common cause of traveler's diarrhea.
B.   *Entamoeba histolytica* is an important cause of dysentery. Amebiasis is often chronic, and the infection can spread to the liver and other organs.

VOCABULARY: TERMS AND DEFINITIONS

The following list contains new terms introduced in this chapter. Use these terms to fill-in the blanks of the sentences that follow and you will have a definition or description of each new term.

Shiga toxin	**viremia**	**cholergen**
bile salts	**villi**	**adhesions**
jaundice	**emulsify**	**infestation**
dysentery	**Peyer's patches**	

1. The inside surface of the intestine is covered with many fingerlike projections called —————————— .

2. —————————— are specialized surface structures on many pathogens that allow them to adhere to the intestinal epithelial cells.

3. The enterotoxin produced by *Shigella dysenteriae* that causes abscesses and ulcerations of the intestinal tract is known as —————————— toxin.

4. —————————— is a yellow color of the skin and eyes caused by bile pigments.

5. The presence of viruses in the blood is called —————————— .

6. —————————— is the living as a parasite on or in a host.

7. The enterotoxin produced by *Vibrio cholerae* is called —————————— .

8. Bloody, pus-filled diarrhea is known as —————————— .

9. Substances in the bile called —————————— act as detergents and help —————————— fats in the intestine.

10. —————————— are collections of lymphoid cells in the intestinal wall that are susceptible to invasion by *Salmonella typhi*.

SELF-TEST OF READING MATERIAL

Match the list of causative agents on the right with the description of the food poisoning on the left.

———— 1. Ergot poisoning resulting from eating bread prepared from grain with fungal growth

———— 2. Toxin produced during sporulation

———— 3. Most severe type of food poisoning following ingestion of preformed toxin

———— 4. Most common type of food poisoning with a heat-stable enterotoxin

———— 5. Mild type of food poisoning associated with rice dishes

———— 6. Acute poisoning capable of producing liver damage and possibly liver cancer

a. *Staphylococcus aureus*
b. *Clostridium perfringens*
c. *Bacillus cereus*
d. *Clostridium botulinum*
e. *Claviceps purpurea*
f. *Aspergillus*

151

7. Botulinum toxin, which causes a fatal food poisoning, is
   1. heat labile
   2. an enterotoxin
   3. heat stable

   4. a protein
   5. a neurotoxin
   6. a lipopolysaccharide

   a. 2,3,6
   b. 1,4,5
   c. 1,2,6

   d. 3,4,5
   e. 1,2,4

8. Infant or intestinal botulism is different from botulism food poisoning because
   a. a neurotoxin is not involved.
   b. infant botulism is due to an organism and not a toxin.
   c. infant botulism results from an infection.
   d. botulism food poisoning is due to an enterotoxin.
   e. it is carried by honey.

9. In the United States, the most common source of botulism is
   a. eclairs.
   b. home canned foods.
   c. fish.

   d. sausages.
   e. water.

10. The recent epidemic of intestinal disease associated with *E. coli* strain O157:H7 was caused by
    a. enterotoxigenic *E. coli*
    b. enterohemorrhagic *E. coli*
    c. enteropathogenic *E. coli*

    d. enteroaggregative *E. coli*
    e. enteroinvasive *E. coli*

11. Pet turtles have been shown to be an important source of
    a. *Salmonella*
    b. *Clostridium botulinum*
    c. *Giardia lamblia*

    d. enterohemorrhagic *E. coli*
    e. *Shigella dysenteriae*

12. The most common cause of cases of post-transfusion hepatitis is
    a. hepatitis A virus
    b. hepatitis B virus
    c. hepatitis C virus

    d. a and b are correct.
    e. All three occur with equal frequency.

13. Which of the following is responsible for most of the waterborne disease epidemics in the United States in which a cause is identified?
    a. hepatitis A virus
    b. enterotoxigenic E. coli
    c. *Shigella dysenteriae*

    d. *Salmonella*
    e. *Giardia lamblia*

14. Suppression of the normal flora of the intestine with antibiotics increases susceptibility to
    a. *Clostridium difficile*
    b. *Salmonella enteritidis*
    c. *Entamoeba histolytica*

    d. a and b are correct.
    e. None of the above.

15. Giardiasis, sometimes called "beaver fever" or "backpacker's disease," was thought to be associated exclusively with drinking water in mountain streams, but is now also found in day care centers associated with diaper changing.
    a. True
    b. False

REVIEW QUESTIONS

1. Various strains of *E. coli* mimic other intestinal infections. List the four main types of virulent strains of *E. coli* along with the disease or organism that each mimics.

2. Compare and contrast the food poisoning toxins of *Staphylococcus aureus* and *Clostridium botulinum*.

3. Why is bile included in some selective microbiological media?

4. What are adhesions and why are they important to intestinal pathogens?

5. What is significant about the infective stages of *Entamoeba histolytica* and *Giardia lamblia*?

6. How do hepatitis A and hepatitis B differ in their transmission?

27 *GENITOURINARY INFECTIONS*

Genitourinary tract infections are caused by a number of opportunists that are part of the normal flora of the body or by pathogens introduced from the outside. While many of the infections are considered sexually transmitted diseases most of the remaining infections occur when there is some breakdown in the barriers that protect the genitourinary tract. In the urinary tract this includes conditions that interfere with the normal flow of urine. Interference with the normal flora in the vagina is a significant first step to the establishment of an infection. Women are also more susceptible to urinary tract infections because of a short urethra and the proximity to the intestinal tract.

KEY CONCEPTS

1. The flushing action of urination is a key defense mechanism against bladder infections.
2. During the childbearing years, a woman's hormones are important in the vagina's resistance to infection because estrogen influences the vagina's normal flora.
3. Transmission of genital tract infections usually requires direct human to human contact.
4. Unsuspected sexually transmitted infections of pregnant mothers pose a serious threat to fetuses and newborn babies.
5. Asymptomatic infections can cause serious genital tract damage and are transmissible to other people.

SUMMARY

I. Anatomy and Physiology
   A. The main structures of the urinary tract, the kidneys, bladder, ureters and urethra, are generally well protected from infection.
   B. Urinary tract infections occur more frequently in women than in men because of the shortness of the female urethra and its closeness to the genital and intestinal tracts.
   C. The vagina is the site of entry for a number of infections

II. Normal Flora Of The Genitourinary Tract
   A. In women, the normal flora is dependent on the action of estrogen hormones that promote the deposition of glycogen in cells lining the vagina. Lactobacilli normally help prevent colonization by pathogens.

III. Urinary Tract Infections
   A. Bladder infections occur when frequent, complete emptying fails to take place. This allows accumulation of urine, a nutritious medium for many bacteria, and permits multiplication of pathogenic bacteria that might be present.
   B. Kidney infection may complicate untreated bladder infection when pathogens ascend the ureters.
   C. Ninety percent of urinary tract infections are caused by enterobacteria from the person's own normal intestinal flora.
   D. *Escherichia coli* is the cause of the majority of urinary tract infections in otherwise healthy women.
   E. *Pseudomonas aeruginosa* infections are difficult to treat because these bacteria are generally resistant to many antibiotics.
   F. Proper collection and culturing of urine specimens is critical in identifying the cause of urinary infections because during urination, normal flora are introduced into the specimen from the lower urethra and vagina.
   G. Causative agents of diseases involving areas of the body other than the urinary tract may be found in urine.

Me

IV. Genital Tract Infections
   A. Bacterial sexually transmitted diseases
      1. Gonorrhea, caused by *Neisseria gonorrhoeae*, is among the most prevalent sexually transmitted diseases. Ophthalmia neonatorum, a very destructive eye disease of the newborn can be acquired during passage through the mother's infected birth canal. Infections in men may extend to the prostate gland and testes. In women, gonococcal infections can spread upward through the uterus and into the fallopian tubes resulting in pelvic inflammatory disease. The resulting scar formation can block the passage of ova through the fallopian tubes, causing sterility or ectopic pregnancy.
      2. Symptoms and complications of chlamydia infections are very similar to those of gonorrhea. Asymptomatic carriers are likewise, common.
      3. Syphilis, caused by *Treponema pallidum*, can imitate many other diseases during its progression. An infected, asymptomatic pregnant woman can transmit the organism to her fetus.
      4. Chancroid is caused by *Haemophilus ducreyi*, an X-factor requiring bacterium. In some tropical countries, it is second only to gonorrhea among the sexually transmitted diseases. Like other ulcerative sexually transmitted diseases, it can increase the risk of acquiring HIV infection.
   B. Viral sexually transmitted diseases
      1. Genital herpes, caused by herpes simplex virus type 2, is among the most common sexually transmitted diseases. Recurrences are common. The viral genome exists in a noninfectious form within nerve cells during asymptomatic periods. However, the disease can be transmitted, indicating that mature virions can be produced. The disease is incurable, but can be suppressed by medication.
      2. Genital warts are caused by human papillomaviruses (HPV). Some HPV strains are strongly associated with cervical cancer. Removal of the warts does not necessarily cure the infection.
      3. HIV infection is currently pandemic and ultimately fatal in most cases. No vaccine or medical cure is yet available, but the spread of infection can be halted by applying current knowledge and practices.
   C. Other genital tract infections
      1. Bacterial vaginosis is the most common cause of vaginal symptoms. It is characterized by a decrease in lactobacilli and an increase in *Gardnerella vaginalis* and anaerobic bacteria.
      2. Vulvovaginal candidiasis is caused by *Candida albicans*, a yeast that is commonly a normal part of the vaginal flora. This organism becomes pathogenic when the normal bacterial flora is reduced by things such as antibiotic treatment.
      3. Vaginitis can be caused by the protozoon *Trichomonas vaginalis*. Infections in men are usually asymptomatic.
      4. Puerperal fever can be caused by *Streptococcus pyogenes* or anaerobic bacteria. Contaminated instruments or hands spread the disease.
      5. Toxic shock syndrome is caused a toxin produced during the growth of certain strains of *Staphylococcus aureus*. This may occur in the vagina after high-absorbency tampons have been left in place too long.

SELF-TEST OF READING MATERIAL

Use the following list of organisms to match the description or name of an infection below.

a. *Haemophilus ducreyi*
b. herpes simplex virus type 2
c. *Trichomonas vaginalis*
d. *Staphylococcus aureus*
e. *Leptospira*
f. *Treponema pallidum*
g. *Neisseria gonorrhoeae*
h. *Pseudomonas aeruginosa*

i. *Candida albicans*
j. human papilloma virus
k. *Chlamydia trachomatis*
l. *Streptococcus faecalis*
m. human immunodeficiency virus
n. *Escherichia coli*
o. herpes simplex virus type 1

_____ 1. Chancroid

_____ 2. Toxic shock syndrome

_____ 3. Genital warts

_____ 4. 90% of urinary tract infections in healthy individuals

_____ 5. Disseminated gonococcal disease

_____ 6. An organism that causes urinary tract infections and is very resistant to antibiotics

_____ 7. Vaginal organism usually kept under control by the normal flora

_____ 8. The "great imitator" disease

_____ 9. Hard chancre

_____ 10. The agent of AIDS

_____ 11. Protozoan infection of women that is usually asymptomatic in men

_____ 12. Treatment for gonorrhea now assumes this organism is also present

_____ 13. The most common gram-positive agent in urinary tract infections

_____ 14. Disease can be transmitted in urine-contaminated water

_____ 15. The "pale thread"

_____ 16. Associated with cancer of the cervix

_____ 17. Infection that mimics gonorrhea

_____ 18. Causes ophthalmia neonatorum

_____ 19. Pelvic inflammatory disease

_____ 20. Causes scarring of fallopian tubes that can lead to ectopic pregnancy

21. The hard chancre of primary syphilis is due to
   a. exotoxins.
   b. inflammatory reaction.
   c. immune complexes.
   d. endotoxins.
   e. Two of the above are correct.

22. Which of the following has a normal resident flora?
   a. lower urethra
   b. upper ureter
   c. kidney
   d. upper urethra
   e. lower ureter

23. The urinary tract is protected from infection by
    a. downward flow of urine.
    b. antimicrobial organic acids in urine.
    c. antibodies in the urine.
    d. All of the above are correct.
    e. Only a and b are correct.

24. The symptoms of secondary syphilis are
    a. due to the presence of large numbers of bacteria.
    b. due to endotoxins.
    c. due to exotoxins.
    d. associated with concurrent gonococcal infections.
    e. due to immune complexes.

25. Genital herpes can be caused by either herpes simplex virus type 1 or type 2, but infections by type 2 virus produces more severe lesions with a greater frequency of recurrence.
    a. True
    b. False

26. *Trichomonas vaginalis* does not have
    a. mitochondria.
    b. a cyst stage.
    c. flagella.
    d. a and b are correct.
    e. a and c are correct.

27. A non-treponemal test for syphilis detects the presence of
    a. treponemal antigens in serum.
    b. beef heart lipids in serum.
    c. antibody-like proteins in serum.
    d. beef heart antibodies in serum.
    e. viable *Treponema pallidum*.

28. Gonorrhea in men is usually asymptomatic and resolves it self without treatment.
    a. True
    b. False

29. Erythromycin is replacing silver nitrate in the prophylaxis of opthalmia neonatorum because silver nitrate is not effective against _____ which also often occurs in combination with gonococcal infections.
    a. *Chlamydia trachomatis*
    b. *Candida albicans*
    e. HIV
    d. *Trichomonas vaginalis*
    e. *Treponema pallidum*

30. Genital warts can be removed with laser treatment or freezing with liquid nitrogen, but this does not cure the infection.
    a. True
    b. False

REVIEW QUESTIONS

1. List the seven symptoms that should alert one to the possibility of a sexually transmitted disease.

2. Describe a non-treponemal test for syphilis and a test that uses treponemal antigen.

3. What is the relation of estrogen to resistance of the vagina to infection?

4. Why are hospital patients and women more likely to a urinary tract infection?

5. Describe the characteristics of the three stages of syphilis?

6. What can account for the recent rapid increase in the number of cases of congenital syphilis.

NOTES

28 *NERVOUS SYSTEM INFECTIONS*

For fairly obvious reasons, infections of the nervous system, especially the central nervous system, are extremely serious. The nervous system does not have any normal flora associated with it. Infectious agents must enter across the blood-brain barrier, by way of the peripheral nerves, or by direct extension from wounds or infections of associated areas such as mastoids, ears or sinuses. Treatment of nervous system infections is often difficult because most drugs or antibiotics will not cross the blood-brain barrier. Representatives of the bacteria, viruses, protozoa and the fungi, nevertheless, manage to establish and produce disease in the nervous system.

KEY CONCEPTS

1. Infectious diseases of the central nervous system are uncommon compared to infections elsewhere in the body because pathogens usually cannot cross the blood-brain barrier.
2. The presence or absence of receptors on the surface of various cells within the nervous system determines their susceptibility to infection and thus the symptoms of an infectious disease.
3. The nervous system is infected primarily from the blood stream, via the nerve cells or through the cranial bone.
4. Antimicrobial medicines are effective in central nervous system infections only if they are able to cross the blood-brain barrier.

SUMMARY

I. Anatomy and Physiology of the Nervous System
    A. The brain and spinal cord make up the central nervous system.
    B. There are two kinds of nerves.
        1. Motor neurons cause parts of body to act.
        2. Sensory neurons transmit sensations such as heat, pain, touch, light, and sound.
    C. Cerebrospinal fluid (CSF) is produced in cavities inside the brain and flows out over the brain, spinal cord and nerves. CSF can be sampled and tested for the cause of central nervous system infection.
    D. Meninges are the three membranes that cover the surface of the brain and spinal cord. Cerebrospinal fluid flows between the two innermost membranes.
    E. Infectious organisms can reach the brain and spinal cord in three ways.
        1. The bloodstream where the blood-brain barrier acts as protection.
        2. The nerves.
        3. Direct extension through the skull by infections in regions such as the middle ear or sinuses.

II. Bacterial Infections of the Nervous System
    A. *Neisseria meningitidis*, the agent of meningococcal meningitis, is a gram-negative diplococcus. Release of endotoxin from the organism can cause shock and death. *N. meningitidis* is capable of causing epidemics.
    B. *Haemophilus influenzae*, a leading cause of bacterial meningitis, is a tiny gram-negative, coccobacillus that requires X- and V-factors for growth. It produces satellites in culture. Some strains have acquired a plasmid that codes for penicillinase.
    C. *Streptococcus pneumoniae*, the agent of pneumococcal meningitis is a gram-positive, encapsulated diplococcus. Pneumonia and ear or sinus infection often accompanies meningitis. *S. pneumoniae* is the most common cause of meningitis in adults.
    D. *Escherichia coli*, an enteric gram-negative rod, is one of leading causes of meningitis in newborns.
    E. *Streptococcus agalactiae*, another leading cause of meningitis of newborns, has a Lancefield Group B cell wall polysaccharide.
    F. *Listeria monocytogenes*, a motile, aerobic, gram-positive rod, causes meningitis in newborns and individuals with compromising conditions such as diabetes.

G. Hansen's disease (leprosy) is characterized by the invasion of peripheral nerves by the acid-fast bacillus, *Mycobacterium leprae*. This organism has not yet been cultivated *in vitro*. Hansen's disease occurs in two main forms, tuberculoid and lepromatous, depending on the immune status of the patient.

III. Viral diseases of the Nervous System
    A. Sporadic encephalitis is most often caused by herpes simplex virus
    B. Epidemic encephalitis in humans is usually caused by arboviruses.
        1. La Crosse encephalitis virus has a transmission cycle that includes *Ades* mosquitoes and rodents.
        2. St. Louis encephalitis virus has a transmission cycle that includes *Culex* mosquitoes and birds.
    C. Poliomyelitis, which involves the motor nerves of both the brain and spinal cord, is caused by three picornaviruses, polioviruses 1, 2, and 3.
        1. Motor nerve destruction leads to paralysis, muscle wasting, and failure of normal bone development.
        2. Post-polio syndrome occurs years after poliomyelitis and is probably caused by the death of nerve cells that had compensated for ones killed by the poliovirus.
    D. Rabies is transmitted mainly through the bite of an infected animal.
        1. The incubation period is long, often measured in months and sometimes years.
        2. The virus multiplies in muscle and then travels through nerve cells to the brain. After multiplying in the brain, it spreads outward, through the nerves, to infect other body tissues, rapidly causing death.

IV. Fungal Diseases of the Nervous System
    A. Cryptococcal meningoencephalitis is an infection of the meninges and brain caused by *Filobasidiella (Cryptococcus) neoformans*. Infection originates in the lung after inhaling dust laden with fungal spores. The organism is associated with old pigeon droppings. This disease is a common, serious complication of AIDS.

V. Protozoan Diseases of the Nervous System
    A. African Sleeping Sickness is caused by *Trypanosoma brucei*.
        1. During infection, the organism shows bursts of growth, each appearing with different surface proteins.
        2. Each of these variants requires that the body respond with a new antibody since the prior one is ineffectual.
    B. Primary amebic meningoencephalitis is caused by *Naegleria fowleri*.
        1. The amoeba enters the nasal cavity and travels along the olfactory nerve tracts to the brain.
        2. Infections are associated with swimming and bathing activities in heated waters.

VOCABULARY: TERMS AND DEFINITIONS

The following list contains new terms introduced in this chapter. Use these terms to fill-in the blanks of the sentences that follow and you will have a definition or description of each new term.

axons	**Negri**	**serum sickness**
meninges	**conjugate**	**meningococcal meningitis**
shock	**cerebrospinal fluid**	**meningitis**
hydrophobia	**petechiae**	

1. A state in which there is insufficient blood pressure to supply adequate flow of blood to vital organs

   is called_____ .

2._____ are long thin extensions of nerve cells that transmit electrical impulses.

3. A clear fluid filling the ventricles of the brain and covering the brain and the spinal cord is called

   the _____ fluid.

4. Characteristic inclusion bodies at the site of rabies virus replication in brain cells

   are called _____ bodies.

5. _____ is an inflammation of the meninges.

6. Hypersensitivity to sera such as horse serum used in rabies treatment is called _____ .

7. _____ literally means "fear of water" but is a painful spasm of the throat and respiratory muscles provoked by swallowing or seeing liquids.

8. Three membranes that cover the surface of the brain and the spinal cord are called the _____ .

9. _____ are purplish spots resulting from small hemorrhages under the skin and are a

   classical sign of _____ .

10. A _____ vaccine is one in which antigenic determinants are covalently linked to another protein for improved immune response.

SELF-TEST OF READING MATERIAL

1. People with tuberculoid leprosy, the least severe form of leprosy, rarely transmit the disease to others.
   a. True
   b. False

2. Which of the following is true about the normal flora of the nervous system?
   a. Only transient organisms are present.
   b. Microorganisms are present only in portions of the central nervous system.
   c. The nervous system does not have a normal flora.
   d. Microorganisms are present only in portions of the peripheral nervous system.
   e. a and c are correct.

3. Infections of the central nervous system and brain are difficult to treat because
   a. drugs cannot pass through the blood-brain barrier
   b. nerve tissue does not regenerate so that healing is very difficult.
   c. most infections are caused by fungi and they are difficult to treat.
   d. drugs also affect the brain and cause damage.
   e. most infections occur only after extensive brain damage.

4. In rabies infections
   a. treatment after infection is possible now because of new viral vaccines.
   b. most cases in the United States today are in wild animals and not dogs.
   c. the incubation period is often long and dependent on the site of the bite and the dose of virus.
   d. a and b are correct.
   e. b and c are correct.

Use the list on the right which describes means of disease transmission to match with the appropriate infection on the left.

_____ 5. poliomyelitis

_____ 6. arbovirus encephalitis

_____ 7. *E. coli* meningitis

_____ 8. pneumococcal meningitis

_____ 9. leprosy

_____ 10. *Listeria* meningitis

_____ 11. cryptococcal meningitis

_____ 12. African sleeping sickness

_____ 13. meningococcal meningitis

_____ 14. rabies

a. inhalation of infectious droplets
b. tsetse fly bite
c. infection in birth canal
d. inhalation of dust particles
e. ingestion of dairy products
f. animal bite
g. direct contact
h. fecal-oral route
i. mosquito bite

15. Hansen's disease or leprosy is best characterized as
    a. a highly contagious disease.
    b. an infection of the peripheral nervous system.
    c. an acute infection.
    d. an infection of the central nervous system.
    e. Two of the above are correct.

16. *Streptococcus pneumoniae*, *Hemophilus influenzae*, and *Neisseria meningitidis* seldom cause meningitis in newborns because
    a. most mothers have antibodies against them and these are transferred across the placenta.
    b. newborns lack the appropriate adhesions for the infections.
    c. most mothers have antigens against them and these are transferred across the placenta.
    d. these organisms lack the appropriate adhesions for the infections.
    e. The statement is false because these organisms do often cause meningitis in newborns.

17. Sentinel chickens are useful in detecting the presence of diseases transmitted by
    a. animal bites.
    b. mosquito bites.
    c. direct contact.
    d. chickens.
    e. aerosol droplet.

18. Rabies can be transferred by corneal transplant.
    a. True
    b. False

19. The causative agent of spongiform encephalopathies appears to be
    a. a virus.
    b. an unusual protein.
    c. common in humans.
    d. a protozoan.
    e. easily cured.

20. Poliomyelitis is actually an infection of the digestive tract that spreads to the nervous system in a small percentage of infected individuals.
    a. True
    b. False

REVIEW QUESTIONS

1. Why is it not a good idea, disease-wise, to squeeze a pimple at the corner of your mouth?

2. Describe how *Naegleria fowleri* gets into the central nervous system and from where.

3. Complete the following table by listing the principal species of bacteria that cause meningitis at the different ages.

AGE GROUP	PRINCIPLE SPECIES THAT CAUSE BACTERIAL MENINGITIS
Newborns	
Children	
Adults	

NOTES

ANSWERS:
    Vocabulary: Terms and Definitions
        1. shock 2. axons 3. cerebrospinal fluid 4. Negri 5. meningitis 6. serum sickness 7. hydrophobia
        8. meninges 9. petechiae/meningococcal meningitis 10. conjugate
    Self-Test of Reading Material
        1. b 2. c 3. a 4. e 5. h 6. i 7. c 8. a 9. g 10. e 11. d 12. b 13. a 14. f 15. b 16. a 17. b 18. a
        19. b 20. a

29 *WOUND INFECTIONS*

A wound or break in the skin surface can permit any number of different pathogens or opportunists to enter the body and initiate colonization, infection and disease. These wounds may be as minor as a friendly scratch from your favorite kitty to as major as someone getting an arm or leg caught in a piece of farm equipment. They all have one important consequence in common; bacteria and fungi can cross our first line of defense against disease. Tetanus results from an organism that remains at the site of infection but causes its destruction with a powerful nerve exotoxin. At the other extreme is gas gangrene that is highly invasive because the organism extends from damaged wound tissue into healthy tissue. This chapter details the major aspects of such infections.

KEY CONCEPTS

1. Bacteria with little invasive ability can produce serious disease if they synthesize toxins that are absorbed and carried to other parts of the body.
2. Because a variety of pathogens may infect wounds, widely differing diagnostic and treatment methods may be required.
3. *Staphylococcus aureus* is the most important cause of wound infections because it is commonly carried by humans.
4. Bacterial exotoxins from different species of bacteria may have the same mode of action but cause distinctly different diseases because the toxins affect different target body cells.
5. The prevalence of certain wound infections is determined by factors such as geography and occupation.
6. Bacteria with little invasive ability when growing alone, can invade tissue when growing with other bacteria.

SUMMARY

I. Features of Wounds
   A. Wound abscesses are localized collections of pus. They do not have a blood supply and inflammation surrounds the abscess. The microorganisms present may not be actively growing and therefore not affected by antibiotics. Microorganisms in abscesses are potential sources of infection elsewhere in the body.
   B. Anaerobic conditions in wounds permit the growth of anaerobic pathogens. Puncture wounds, dirty wounds, and those with dead tissue are likely to be anaerobic.
   C. Even microorganisms usually considered harmless, can cause serious infections if the tissue is crushed or the host is immunodeficient.

II. Bacterial Infections of Wounds
   A. *Staphylococcus aureus* (coagulase-positive staphylococcus) is the most common cause of infections in surgical wounds.
   B. Some strains of Streptococcus pyogenes can destroy tissue and cause shock.
   C. *Pseudomonas aeruginosa* is a common cause of burn infections. It is usually resistant to most antibiotics.
   D. Clostridial wound infections
      1. Tetanus (lockjaw) is caused by a toxin produced by *Clostridium tetani*. Colonization, even of minor wounds, by *C. tetani* can result in tetanus. The disease is often fatal, but can be prevented by proper immunization and care of all wounds.
      2. Gas gangrene is marked by necrosis of muscle tissue. It is most commonly caused by the anaerobe *Clostridium perfringens*.
   D. Nonclostridial anaerobic infections of wounds
      1. The primary causative agents are *Bacteroides*, *Fusobacterium* and *Peptostreptococcus*.
      2. Actinomycosis is caused by *Actinomyces israelii*, which is considered normal flora of the mouth and intestinal tract. Infections can originate from dental and intestinal surgery.

    E. Bite wounds
1. Infection of bite wounds may result from a single species of bacterium or from multiple species acting synergistically.
2. *Pasteurella multocida* infection commonly complicates animal bite wounds.
3. Cat scratch disease is characterized by localized skin papules, followed by large and often pus-filled lymph nodes .
4. Rat bite fever can be caused by either *Spirillum minus* or *Streptobacillus moniliformis*. *S. minus* cannot be cultivated in vitro while *S. moniliformis* characteristically has cell wall deficient variants called L-forms.
5. Human bites often result in severe synergistic infections. Prompt treatment is important.

III. Fungal Infections of Wounds
    A. Sporotrichosis is a chronic fungal disease mainly of those who work with vegetation. The causative agent is *Sporothrix schenckii*, usually introduced into wounds caused by thorns and splinters.

SELF-TEST OF READING MATERIAL

Match the organism on the right with its description on the left.

_____ 1. The major cause of death in burn patients

_____ 2. The major cause of surgical wound infections

_____ 3. Organism spreads to healthy tissue with alpha toxin

_____ 4. Introduced into wounds by thorns and splinters

_____ 5. Produces a greenish discoloration of wound

_____ 6. A non-invasive exotoxin producer

_____ 7. The most common cause of burn infections

_____ 8. Produce "sulfur granules" in wounds

_____ 9. Causes cat scratch fever

_____ 10. Hospital environment is source for patient

_____ 11. Most common cause of myonecrosis

_____ 12. Newly recognized causes of wound infections

_____ 13. Most difficult burn infection to treat

_____ 14. Hyperbaric oxygen treatment sometimes helpful

_____ 15. A filamentous bacterium

_____ 16. Patient's own body is most common source

a. *Staphylococcus aureus*
b. *Legionella*
c. *Rhodococcus*
d. *Pseudomonas aeruginosa*
e. *Clostridium tetani*
f. *Clostridium perfringens*
g. *Actinomyces israelii*
h. *Pasteurella multocida*
i. *Rochalimea henselae*
j. *Spirillum minus*
k. *Sporothrix schenckii*

_____ 17. Causes rat bite fever

_____ 18. Soil is most common source

_____ 19. Results from animal bites

_____ 20. Organism grows in dead tissue

REVIEW QUESTIONS

1. Describe three reasons microorganisms in abscesses are not always killed even when antibiotics are used.

2. How does the presence of sutures affect the infecting dose of *Staphylococcus aureus* required to form an abscess?

3. What four factors are responsible for a shift in the spectrum of bacteria that cause surgical wound infections? Explain.

4. What two conditions contribute to the development of gas gangrene?

5. Why is *Pseudomonas* such a feared organism in burn wards?

6. Why is actinomycosis not a mycosis?

7. Complete the following table by listing the organisms usually associated with the type of wound.

TYPE OF WOUND	CAUSATIVE AGENT
Trauma wound	
Surgical wound	
Burn	
Animal bite	

30 *BLOOD & LYMPHATIC INFECTIONS*

The blood and lymph systems are characteristically sterile. The introduction of infectious agents has serious consequences because the two systems can carry the organisms throughout the body and produce systemic infections. The blood and lymph have some natural resistance to infectious agents but when this resistance is overcome, serious disease results. Bacteremia is the simple presence of bacteria in the blood while septicemia usually implies the actual growth of bacteria in the bloodstream. Infections range from chronic, such as subacute bacterial endocarditis, to acute as in the plague. Viral and protozoan infections of the blood and lymph are serious diseases that affect many millions of people worldwide. Malaria still remains the greatest killer of mankind in spite of the successes of the 1950's in controlling the mosquito vector of this devastating disease. More than a million people a year still die from malaria. The difficulty with viral infections of the blood and lymph is the same as for any viral infection; the lack of effective means of treatment. This is no more painfully evident than in HIV infections that ultimately result in AIDS.

KEY CONCEPTS

1. A systemic infection represents failure of the body's mechanisms for keeping infections localized to one area.
2. The inflammatory response, although vitally important in localizing infections, can be life threatening if generalized.
3. Systemic infections threaten the transportation of oxygen and nutrients to body tissues and removal of waste products.
4. The circulatory system can expose all the body's tissues to infectious agents and their toxins.

SUMMARY

I. Anatomy and Physiology of the Circulatory and Lymphatic Systems
    A. The blood vascular system includes the heart, arteries, veins and capillaries.
    B. The lymphatic system consists of the lymphatic vessels and lymph nodes. The blind-ended lymph vessels collect extracellular fluid and carry it back to the bloodstream.

II. Bacterial Diseases of the Blood Vascular System
    A. Subacute bacterial endocarditis (SBE) involves the valves and lining of the heart chambers and is commonly caused by oral streptococci or *Staphylococcus epidermidis*. Infection usually begins on valves previously damaged by disease, or on congenital abnormalities.
    B. Acute bacterial endocarditis is usually caused when virulent bacteria such as *Staphylococcus aureus* or *Streptococcus pneumoniae* enter the bloodstream from a focus of infection elsewhere in the body, or when contaminated material is injected by drug abusers. The normal heart is commonly infected.
    C. Septicemia usually results from bacteremia. It is often caused by gram-negative organisms, mainly enterobacteria and anaerobes of the genus *Bacteroides*.

III. Bacterial Diseases Involving the Lymph Nodes and Spleen
    A. The mononuclear phagocyte system is involved in these diseases.
    B. Tularemia is often transmitted from wild animals to humans by insects and ticks. The cause is the gram-negative aerobe, *Francisella tularensis*.
    C. Brucellosis is usually acquired from cattle or other domestic animals and is caused by species of *Brucella*. Pasteurization of milk is an important control measure.
    D. Plague, once pandemic, is now endemic in rodent populations. It is caused by *Yersinia pestis* and is transmitted to humans by fleas or by persons suffering from pneumonic plague.

IV. Viral Diseases of the Lymphoid System
   A. Infectious mononucleosis is a viral disease principally of B lymphocytes. The causative agent is the Epstein-Barr virus which establishes a lifelong latent infection of lymphocytes, making them "immortal." The virus may be partly responsible for Burkitt's lymphoma and nasopharyngeal carcinoma.
   B. The acquired immune deficiency syndrome (AIDS) is a late manifestation of human immunodeficiency virus (HIV) infection. Relentless destruction of helper lymphocytes over months or years leads to AIDS.
   C. Yellow fever is caused by an arbovirus and is characterized by fever, jaundice and hemorrhage.

V. Protozoan Disease of the Blood
   A. Malaria is caused by four species of *Plasmodium* and is transmitted by the bite of the *Anopheles* mosquito.

SELF-TEST OF READING MATERIAL

Use the list of diseases below to match their appropriate description.

a. subacute bacteria endocarditis
b. acute bacterial endocarditis
c. septicemia
d. tularemia
e. brucellosis
f. plague
g. infectious mononucleosis
h. acquired immunodeficiency syndrome
i. yellow fever
j. malaria

_____ 1. *Streptococcus epidermidis* and alpha-hemolytic streps are causative agents.

_____ 2. It is endemic in rodent populations worldwide.

_____ 3. It is an occasional complication of infectious disease with bacteremia.

_____ 4. Name means "bad air."

_____ 5. Agent can be isolated from Burkitt's lymphoma.

_____ 6. Enlarged lymph nodes called bubos are a hallmark.

_____ 7. Fleas serve as vectors.

_____ 8. The causative agents are the body's normal flora.

_____ 9. *S. aureus* and *Streptococcus pneumoniae* are the causative agents.

_____ 10. Shock is common.

_____ 11. Transmission by tick bites, inhalation, and skinning wild animals are possible.

_____ 12. Infection has a high incidence in 15-24 year olds.

_____ 13. Recurrent bouts of fever are a sign of disease.

_____ 14. Saliva contains virus but is not significant in transmission.

_____ 15. Individuals with sickle cell anemia are resistant to infection.

_____ 16. This is a zoonosis.

_____ 17. This condition is most severe with gram-negative bacteria.

_____ 18. Infected B-lymphocytes can reproduce in culture indefinitely.

_____ 19. This infection has a mosquito vector.

_____ 20. This infection produces a steep-walled ulcer where it enters the skin.

_____ 21. This agent is commonly introduced into circulation by dental procedures.

_____ 22. Black Death is another name for this disease.

_____ 23. Pneumonic form is transmitted by aerosols.

_____ 24. It commonly occurs in congenitally or disease damaged hearts.

_____ 25. Pneumonia from *Pneumocystis carinii* is the cause of death in more than half of the cases.

_____ 26. Disease causes more than a million deaths annually.

_____ 27. These infect erythrocytes.

_____ 28. This is often seen as complication of intravenous drug use.

_____ 29. During this infection, there is a gradual loss of T-4 helper cells over time.

_____ 30. Kissing disease is another name.

_____ 31. T-4 helper cells are targets.

_____ 32. This is a retrovirus.

REVIEW QUESTIONS

1. What types of antimicrobial agents are carried by the lymph?

2. Contrast acute and subacute bacterial endocarditis.

3. Describe the relationship between tumor necrosis factor and shock.

4. What is disseminated intravascular coagulation and how is it initiated?

5. Describe the changes to *Yersinia pestis* when it grows in a human host.

6. Compare and contrast bacteremia and septicemia.

7. Make a list of the behaviors of people that put them at risk for HIV infection.

8. Diagram the life cycle of *Plasmodium malariae*.

9. What is the difference relative to transmission between bubonic and pneumonic plague?

10. A man in New Mexico contracted plague and died before it could be determined where he got the infection. According to his family, his only recent contact with any kind of animal occurred when he crawled under his house to remove a neighbor's puppy that had buried something and would not come out. How would guess he got plague.

31 *HIV DISEASE AND COMPLICATIONS OF IMMUNODEFICIENCY*

Acquired immunodeficiency syndrome (AIDS) is a late manifestation of human immunodeficiency virus (HIV) infection.. This chapter describes the HIV virus, transmission and infection. The HIV disease is not highly contagious so that HIV infection and AIDS are preventable. AIDS is actually the result of the failure of the immune system to protect against pathogens and opportunists. Many of the things that cause death in AIDS patients are harmless to someone with an intact immune system. This chapter describes the primary causes of infection and death in AIDS patients.

KEY CONCEPTS

1. The signs and symptoms of a person with AIDS are due mainly to the opportunistic infections and tumors that complicate HIV disease.
2. HIV is not highly contagious.
3. The AIDS pandemic could be stopped by changes in human behavior.

SUMMARY

I. Introduction
   A. AIDS was first recognized because a group of previously healthy young men developed life-threatening infections from organisms generally considered harmless.
   B. Worldwide, it is estimated that a new infection with an AIDS-causing virus occurs about every 13 seconds.

II. AIDS and HIV
   A. Acquired immunodeficiency syndrome (AIDS) is a late manifestation of human immunodeficiency virus (HIV) disease.
      1. Symptoms of HIV disease
         a. "Flu-like" symptoms can occur six days to six weeks after infection by the virus.
         b. Asymptomatic persons can transmit the disease.
         c. The disease progresses in the absence of symptoms.
         d. The onset of immunodeficiency may be marked by LAV, ARC, tumors, or opportunistic infections.
         e. Hairy leukoplakia is probably the result of Epstein-Barr virus reactivation.
      2. Human immunodeficiency virus, type 1 (HIV-1) is the principle cause of AIDS in the United States. It is a single-stranded RNA virus belonging to the Retrovirus family. There are nine subtypes, A through I, based on distinctive nucleic acid sequences. It is easily inactivated by many disinfectants.
         a. Important structural components include the gp120 (SU) surface antigen, the gp41 (TM) transmembrane antigen, the p17 (MA) matrix protein, and the p25 (CA) capsid antigen
         b. The virion contains three important enzymes, reverse transcriptase (RT), protease (PR), and integrase (IN).
         c. There are two copies of the viral genome.
         d. Two genes, gag and pol are translated as a unit, the resulting large protein then splits into PR, RT, IN, CA, MA, and other viral proteins.
         e. The env gene is translated from a spliced messenger RNA, the resulting protein being processed by host cell enzymes to give SU and TM.
         f. Six additional genes are mainly responsible for the regulation of viral reproduction.

3. The disease is spread primarily by sexual intercourse, by blood, and from mother to fetus or newborn.
    a. Early on, the epidemic involved mainly promiscuous homosexual men. Transmission has slowed among this group, but it still remains a leading source of new infections.
    b. Abusers of intravenous drugs who share needles represent an important factor in the epidemic because they spread HIV by both blood and acts of prostitution. There has been little response to control efforts.
    c. Blood transfusions and blood products are now relatively minor modes of transmission because donated blood is screened and recombinant clotting factor is available for hemophiliacs.
    d. In the United States, AIDS is the leading cause of death in the 25-44 year age group. The percentage of cases due to heterosexual transmission is increasing, as is the percentage of cases in women and children.
4. A number of different kinds of body cells can be infected by HIV, including those of the brain, intestinal epithelium, T-4 lymphocytes, and macrophages.
    a. In T-4 lymphocytes, macrophages, and some other cells, the CD-4 surface antigen is an important receptor for the virus, binding to its gp120 (SU) antigen.
    b. Entry of the virus into T lymphocytes requires the virus attach specifically to a chemokine receptor of the lymphocyte, CXCR4 (formerly called fusin).
    c. Entry into macrophages requires the virus attach to a different chemokine receptor, CCR-5.
    d. Inside the host cell, a DNA copy of the viral genome is made through the action of reverse transcriptase, a complementary DNA strand is made, and the double-stranded DNA is inserted into the host genome as a provirus by viral integrase (IN).
    e. Replication of mature virions results in the death of infected lymphocytes, where as macrophages can release infectious HIV without cell death.
    f. HIV mutants readily appear because reverse transcriptase (RT) is prone to make errors in copying the viral genome.
    g. Mechanisms of T-4 lymphocyte destruction include viral replication, accumulation of viral products, host immune attack, and fusion of cells with lysis of the resulting syncytium.
    h. Following infection, large amounts of HIV are released into the blood stream whether or not symptoms occur. Thus tissues throughout the body become infected. There is a transitory drop in the number of CD4+ lymphocytes.
    i. The immune system largely clears infectious virus from the blood stream, and CD4+ lymphocytes return to nearly normal levels.
    j. The battle between the immune system and the virus continues, with huge numbers of CD4+ lymphocytes and virions destroyed each day. The immune system gradually falls behind, finally becoming unable to limit the infection. Viremia recurs and AIDS develops in half the cases in 9-10 years, longer in the rest.
5. Prevention of AIDS depends on changes in human behavior. HIV disease is not highly contagious.
    a. Persons unsure whether they might be infected with HIV are advised to get a blood test.
    b. Educational techniques must be designed specifically for each group at high risk of infection.
    c. It is unlikely that an effective vaccine will become available in the near future, although a number of possibilities are being examined.
6. Treatment of AIDS depends on antiviral medications and prevention and treatment of opportunistic infections and tumors.
    a. Combinations of different antiviral medications in conjunction with therapy directed at infectious complications, has significantly improved the life expectancy and quality of life of AIDS patients.
    b. The effectiveness of treatment is limited by expense and side effects of medications, and thew development of resistance to the medications By HIV and other infectious agents.

III. Malignant tumors that complicate acquired immune deficiencies mostly fall into three types, Kaposi's sarcoma, lymphomas, and carcinomas arising from the epithelium of the anus or uterine cervix. Viruses are suspected to be causative.
   A. Kaposi's sarcoma is a malignant tumor arising from blood or lymphatic vessels.
      1. The tumor is a complication of organ transplantation, AIDS and other acquired immune deficiencies.
      2. It is common enough to be one of the surveillance conditions for AIDS although it may appear before severe immune deficiency develops.
      3. Accumulating evidence implicates a previously unknown herpes virus as a causative agent.
   B. Lymphomas are malignant tumors that arise from lymphoid cells. They have a markedly increased incidence in AIDS and transplant patients.
      1. Both B and T lymphocytes can give rise to lymphomas but B cell lymphomas are more common in patients with immune deficiency.
         a. Strong evidence exists that Epstein-Barr virus plays a causative role in these tumors.
   C. There is an increased rate of anal and cervical carcinoma in people with HIV disease.
      1. These tumors arise from epithelial cells and therefore differ from Kaposi's sarcoma and lymphomas.
      2. Cancers of the uterine cervix in women and of the anus in women and homosexual men are strongly associated with human papillomavirus (HPV) types 16 and 18.
         a. HPV types 16 and 18 are transmitted by sexual intercourse.
         b. Replication of the viruses increases in the presence of immune deficiency.
         c. Precancerous changes are detectable before the development of cancer.

IV. Immunodeficient individuals are susceptible to the same infectious diseases as other people. They also contract severe diseases from agents that are harmless to people with an intact immune system.
   A. Pneumocystosis is a serious lung disease first recognized in malnourished premature infants.
      1. The disease symptoms develop slowly, with gradually increasing shortness of breath and rapid breathing. Fever and cough may or may not be present. Patients can die from lack of oxygen.
      2. The causative agent is *Pneumocystis carinii*, a tiny unclassified organism, most likely a fungus.
      3. Experimentally, the tiny spores are inhaled into the lungs, attach to the air sacks, which will become filled with fluid, the multiplying *P. carinii*, and mononuclear cells. The alveolar walls thicken, impeding the passage of oxygen into the blood.
      4. The organisms occur commonly in humans and other animals, generally causing asymptomatic infections that become latent. Reactivation of latent infections is probably the source of *P. carinii* in most immunodeficient patients, but environmental sources or person-to-person transmission is probable in some cases.
      5. The disease can be prevented in most cases by giving trimethoprim-sulfamethoxazole by mouth as soon as evidence of immunodeficiency develops. An inhaled medication, pentamidine, is also effective.
      6. Trimethoprim-sulfamethoxazole and other medications are available for treatment of pneumocystosis, along with oxygen therapy. For unknown reasons, patients with HIV disease develop intolerance to trimethoprim-sulfamethoxazole more often than other patients.
   B. Toxoplasmosis is rare among immunocompetent individuals, but can be a serious problem for those with malignant tumors, organ transplants, and HIV disease. Toxoplasmosis can also be congenital.
      1. Symptoms of toxoplasmosis differ for the immunologically competent, the fetus, and the immunodeficient.
         a. Infection is common in uncompromised individuals but only about 10% develop symptoms. Typically symptoms consist of a sore throat, fever, enlarged lymph nodes and spleen, sometimes with a rash. The clinical picture can be confused with infectious mononucleosis. Symptoms disappear over weeks or months and generally don't need treatment.
         b. Infection of the fetus occurs about 50% of the time when the mother develops symptomatic or asymptomatic infection during pregnancy. Early in pregnancy, miscarriage can occur, or the baby is born with birth defects or damage to the brain or other organs. Infections later in pregnancy are usually milder, but can result in epilepsy, mental retardation, or recurrent retinitis in the child.
         c. Toxoplasmosis is common and life threatening in those with immunodeficiencies. More than half the cases develop encephalitis, and death often occurs because of involvement of the brain, heart and other organs.

2. The causative agent is *Toxoplasma gondii*, a tiny banana-shaped protozoan that undergoes sexual reproduction in the intestinal epithelium of cats but can infect humans and many other vertebrates.
    a. Oocysts excreted in the feces of acutely infected cats become infectious in soil and contaminate food, water, and fingers. The flesh of animals that have ingested oocysts contains pseudocysts filled with *T. gondii*, also infectious for humans and other carnivores.
3. An enzyme produced by *T. gondii* aids penetration of the host cell membrane. Proliferation of the protozoan results in destruction of the host cell and release of the microorganisms to infect other cells. Normally, host immunity brings the process under control and *T. gondii* is confined to pseudocysts.
    a. With immunodeficiency, the protozoans are released from the pseudocysts and cause a generalized infection.
4. *Toxoplasma gondii* is present worldwide. Most people become infected by oocysts from cats. Epidemics have resulted from contaminated drinking water. Eating rare meat, especially pork and lamb, can also cause infection from pseudocysts.
5. Preventive measures include handwashing after contact with soil, cat litter, or raw meat; washing fruits and vegetables; thorough cooking of meats; preventing cats from eating birds and rodents.
    a. Patients with HIV disease and those to receive immunosuppressive therapy are tested for antibody to *T. gondii*. A positive test indicates the presence of pseudocysts in their body tissues. Prophylactic medication is given when their CD4+ cells decline to low levels. Trimethoprim-sulfamethoxazole is effective.
C. Cytomegalovirus (CMV) is a common cause of blindness in AIDS patients. Like other herpesviruses, it is often acquired early in life, then becomes latent, only to reactivate with development of immunodeficiency.
    1. As with toxoplasmosis, symptoms of CMV differ in immunocompetent individuals, fetuses, and those with acquired immunodeficiency.
        a. Immunocompetent individuals generally have asymptomatic infections; fetuses may develop cytomegaloinclusion disease, or appear normal at birth but show mental retardation or hearing loss later; immunodeficient individuals develop fever, gastrointestinal bleeding, mental dullness, and blindness.
    2. Cytomegalovirus is an enveloped, double-stranded DNA virus. Infected cells are enlarged and have an "owl's eye" appearance because of a large intranuclear inclusion body surrounded by a clear halo. Multiple strains can be identified by their endonuclease digests.
    3. A wide variety of tissues are susceptible to infection. The virus can exist in latent form, a slowly replicating form, or a fully replicating form. Co-infection with HIV results in a fully productive infection and tissue death.
    4. The virus occurs worldwide, and most adults have been infected. Infants with CMV secrete the virus in saliva and urine for months or years. The virus may also be found in semen and cervical secretions, and is commonly transmitted sexually. Breast milk may also contain the virus.
    5. No vaccine is available. Condoms decrease the risk of sexual transmission. Those with immunodeficiency are advised to avoid daycare centers, and to wash their hands after contact with saliva, urine or feces. Tissue donors are screened for antibody to CMV, and if positive, their tissues are not used for persons negative for the antibody. The antiviral medication, ganciclovir, can be given to prevent CMV retinitis.
    6. Ganciclovir and foscarnet are helpful in treating CMV disease, but can cause serious side effects.
D. Mycobacteria commonly cause severe infections in immunodeficient patients. *Mycobacterium tuberculosis* and *Mycobacterium avium* complex (MAC) organisms are most commonly responsible.
    1. Immunocompetent people usually have asymptomatic or mild infections with MAC organisms, but immunodeficient patients may have fever, drenching sweats, severe weight loss, diarrhea and abdominal pain.
    2. MAC is an acronym for *Mycobacterium avium* complex, comprised of *M. avium* and *Mycobacterium intracellulare* strains. They can be distinguished from *M. tuberculosis* by biochemical tests and nucleic acid probes.
    3. MAC organisms enter the body via the lungs and gastrointestinal tract. They are taken up by macrophages but resist destruction and are carried to all parts of the body. In immunodeficiency, MAC multiply without restriction, producing massive numbers of organisms in blood, intestinal epithelium and tissues. There is little or no inflammatory response to the bacteria.

4. MAC bacteria may be present in water, food, and dust. In AIDS patients they are the most common cause of generalized bacterial infection.
5. No generally effective measures are available for preventing exposure to MAC bacteria. Prophylactic antimycobacterial medication is advised for severely immunodeficient patients, but it can fail to prevent infection.
6. Treatment requires two or more antimycobacterial medications used together for the life of the patient.

SELF-TEST OF READING MATERIAL

1. Which of the following limit(s) anti-HIV therapy?
   a. cost
   b. toxicity
   c. viral resistance
   d. a and c
   e. a, b, and c

2. The variability in HIV virions is due to frequent errors made by reverse transcriptase copying the viral genome.
   a. True
   b. False

3. *Toxoplasma gondii*
   1. is a virus.
   2. reproduces sexually in cats.
   3. reproduces asexually in cats.
   4. is a protozoan.
   5. reproduces sexually in humans.
   6. reproduces asexually in humans.
   7. is a fungus.
   8. infects a variety of animals.
   9. is transmitted by a mosquito.

   a. 1,2,5,8
   b. 2,3,4,6,8
   c. 2,3,4,9
   d. 3,4,5,8
   e. 2,6,7,9

4. The incubation period for HIV is
   a. 10 years.
   b. 2 to 6 months.
   c. 6 days to 6 weeks.
   d. 2-3 weeks.
   e. 2-3 days.

5. HIV infection is preventable.
   a. True

6. Symptoms of HIV infections are often
   a. asymptomatic.
   b. mild and attributed to the "flu."
   c. severe but subside within six weeks.
   d. Kaposi's sarcoma
   e. Two of the above are correct.

7. The AIDS-related complex (ARC) is a group of infections that resemble AIDS.
   a. True
   b. False

8. Subtypes of HIV-1, the causative agent of HIV infection and AIDS vary in their
   a. geographical distribution.
   b. ability to infect by different routes.
   c. host species.
   d a and b are correct.
   e. a and c are correct.

9. HIV is _____ contagious than childhood chickenpox.
   a. much more
   b. about as
   c. much less

10. It is now established that HIV-1 was transmitted to humans through contact with the blood of a mangabey.
    a. True
    b. False

11. The AIDS epidemic is increasing most rapidly among
    a. homosexual males.
    b. heterosexual contacts.
    c. intravenous drug abusers.
    d. a and c are correct.
    e. b and c are correct

12. The effectiveness of treatment of HIV is limited by
    a. cost.
    b. side effects of medications.
    c. development of resistance by virus.
    d. All of the above are correct.
    e. a and b are correct.

13. Kaposi's sarcoma is a malignant tumor unique to patients with AIDS.
    a. True
    b. False

14. Most AIDS victims die from
    a. pneumocystosis.
    b. toxoplasmosis.
    c. tuberculosis.
    d. Kaposi's sarcoma.
    e. lymphoma.

15. Cytomegalovirus is a common cause of which of the following in AIDS?
    a dementia
    b. tuberculosis
    c. blindness
    d. lymphoma
    e. Kaposi's sarcoma

REVIEW QUESTIONS

1. Why should AIDS patients avoid day care centers?

2. What symptoms define the AIDS related complex (ARC)?

3. List five AIDS defining conditions used for surveillance. How many of these are opportunistic infections?

4. List 10 ways to eliminate or decrease the risk of HIV infection.

Answers
  Self-Test
    1. e 2. a 3. b 4. c 5. a 6. e 7. b 8. d 9. c 10. b 11. e 12. d 13. b 14. a 15. c

NOTES

32 ENVIRONMENTAL MICROBIOLOGY

We usually think of dirt as being a place of filth and infection; and it does contain a number of pathogenic organisms. However, the microbial populations of the soil are staggering both in number and in variety. The nature of the soil is dependent on its microbial populations. Microorganisms in the soil provide nutrients in their proper form to plants by processes ranging from nitrogen fixation to decomposition. A number of conditions of the microenvironment, in turn, influence the growth and well-being of the microbial population. Soil is a dynamic, living community. Microorganisms are even responsible for the sweet, fresh smell of the soil that we so appreciate in the spring after a very hard, sterile winter.

KEY CONCEPTS

1. The environment immediately surrounding a microorganism, the microenvironment, is most important.
2. Different microorganisms are capable of growing in a wide variety of environments but each will grow best in those environments for which it is well adapted.
3. Life supported by solar energy and photosynthetic primary production is very important, but life sustained by chemosynthetic primary production may actually be more common.
4. Microorganisms are essential for recycling the biologically important elements oxygen, carbon, nitrogen, sulfur, and phosphorus.
5. Microorganisms play important roles in bioremediation, the biological cleanup of pollution.

SUMMARY

I. Principles of Microbial Ecology
   A. Microbial competition demands rapid reproduction and efficient use of nutrients.
   B. Microbial populations both cause and adapt to environmental changes.
   C. Some bacteria can grow in a low nutrient environment.

II. Microorganisms and Soil
   A. Soil teems with a variety of life forms including bacteria, algae, fungi, protozoa, nematodes, worms and other eukaryotic organisms.
   B. Environmental influences on the bacterial and fungal flora in soil.
      1. Moisture in the soil affects the oxygen supply. Waterlogged soils primarily support anaerobic organisms while relatively few organisms grow in desert soils.
      2. Highly acid conditions inhibit the growth of most bacteria. Fungi are generally more tolerant of both high and low pH.
      3. Temperature regulates microbial activity.
      4. Availability of nutrients limits the size of the microbial population.

III. Energy Sources for Ecosystems
   A. Until recently, it was thought that all life depended on photosynthetic organisms for their ultimate food supply.
   B. The discovery of extremophiles has changed that thought.

IV. Aquatic Environment
   A. Water has several unique properties that make it so essential for all living cells.
   B. Specialized aquatic habitats include salt lakes and iron-containing springs.

V. Biochemical Cycling
   A. Oxygen cycles between respirers and photosynthesizers.
   B. Carbon cycles between organic compounds and carbon dioxide.
      1. Producers fix atmospheric CO_2 into organic compounds.
      2. Decomposers use organic material as a source of carbon for various biological processes.
   C. Nitrogen cycles between organic compounds and a variety of inorganic compounds.
      1. Ammonification involves the degradation of cell components such as proteins to CO_2, ammonium and sulfate ions and water through a series of steps.
      2. Nitrification is the microbial oxidation whereby ammonium is converted to nitrite and then to nitrate.
      3. Denitrification occurs when some anaerobic bacteria use nitrate as a final electron acceptor resulting in the liberation of N_2 and the loss of nitrogen to the ecosystem.
      4. Nitrogen fixation involves prokaryotic microorganisms that are capable of reducing nitrogen gas to ammonium ion that is then used as the amino group of amino acids
         a. Nitrogen fixation is accomplished by two types of organisms.
         b. Free-living nitrogen fixers
         c. Symbiotic nitrogen fixers
         d. How roots are infected with rhizobia.
   D. Phosphorus also cycles between organic and inorganic forms.
   E. The sulfur cycle involves several important types of organisms.

VI. Bioremediation: The Biological Cleanup of Pollutants
   A. Pollutants.
   B. Means of bioremediation

VOCABULARY: TERMS AND DEFINITIONS

The following list contains new terms introduced in this chapter. Use these terms to fill-in the blanks of the sentences that follow and you will have a definition or description of each new term.

denitrification	mesophiles	humus
producers	immobilization	putrefaction
nitrogen fixation	microenvironment	mineralization
decomposers	ecosystem	consumers
geosmins	oligotrophic	

1. Bacteria that grow best at temperatures between $20^{\circ}C$ and $50^{\circ}C$ are called _____ .

2. _____ is the chemical transformation of nitrogen from the atmosphere into a form that plants can use.

3. An _____ environment has a low nutrient content.

4. _____ is the conversion of an element from an inorganic form to an organic form.

5. Bacterial metabolites which give soil its characteristic odor are called _____ .

6. The environment immediately surrounding an individual cell is known as the _____ .

7. The community of living organisms in a given area that are interacting with the nonliving environment of that area,

   form an _____ .

8. _____ are organisms that convert carbon dioxide into organic matter.

9. Organisms that metabolize organic matter synthesized by the _____ are

   called _____ .

10. The _____ digest and convert dead plant and animal material into small molecules that can

   be used by both the _____ and the _____ .

11. _____ is the loss of nitrogen from the soil when bacteria convert nitrate to gaseous nitrogen.

12. _____ is the conversion of an element from an organic form into an inorganic form.

13. The organic material in soil derived from the decomposition of plant, animal and microbial matter

   is called _____ .

14. _____ is the breakdown of protein that results in a very odorous product.

SELF-TEST OF READING MATERIAL

1. Phosphatase is
   a. a mineral.
   b. a carbonated fountain drink.
   c. an enzyme that causes the mineralization of organic phosphate.
   d. an enzyme that causes the mineralization of inorganic phosphate.
   e. an enzyme that causes the immobilization of inorganic phosphate.

2. The largest group of bacteria in the soil are
   a. actinomycetes.
   b. *Clostridium*.
   c. fungi.
   d. *Bacillus*.
   e. myxobacteria.

3. Nitrogen fixation is brought about by two types of microorganisms,
   a. actinomycetes and fungi.
   b. free-living and symbiotic.
   c. *Bacillus* and *Clostridium*.
   d. parasitic and symbiotic.
   e. nitrifiers and denitrifiers.

4. The microorganisms associated with the sulfur cycle are
   a. yellow-pigmented.
   b. heterotrophs.
   c. actinomycetes.
   d. autotrophs.
   e. aerobic.

5. Because bacteria are inhibited under highly acid conditions, acid soils will be dominated by
   a. fungi.
   b. protozoa.
   c. nitrogen fixers.
   d. actinomycetes
   e. Acid soil does not have any significant microbial populations.

185

REVIEW QUESTIONS

1. Describe the succession of organisms in a carton of milk that has not been refrigerated. Is this all bad? Explain.

2. Why are herbicides and pesticides so harmful to the ecology of the soil?

3. What conditions affect the growth of microorganisms in the soil?

4. Outline the steps of one of the geochemical cycles.

5. What is the microenvironment of a cell? Why is it important?

ANSWERS:

Vocabulary: Terms and Definitions

1. mesophiles 2. nitrogen fixation 3. oligotrophic 4. immobilization 5. geosmins 6. microenvironment
7. ecosystem 8. producers 9. producers/consumers 10. decomposers/producers/consumers 11. denitrification
12. mineralization 13. humus 14. putrefaction

Self-Test of Reading Material

1. c 2. a 3. b 4. d 5. a

NOTES

33 *WATER AND WASTE TREATMENT*

The treatment of our household and industrial waste is a problem that literally grows larger each day. The not too dissimilar threat of cholera was the driving force behind the development of sewage systems in the last century. Today our environment and, possibly again, our health is being threatened by the wastes that we are producing. We are developing new solutions to help solve these problems but they at first appear expensive until we seriously consider the cost if we do nothing. The good news is that individuals can do something to help. For example, composting by individuals will remove some 20-30% of the burden of solid waste disposal.

KEY CONCEPTS

1. Waterborne diseases result from contamination of water with a variety of microorganisms and chemicals.
2. Adequate water treatment and regular testing assure safe drinking water and recreational waters.
3. Proper sewage treatment, necessary to ensure the health of the community, depends on the stabilization of wastes by microorganisms.
4. Pathogenic bacteria are usually eliminated by the secondary sewage treatment process.
5. Methods such as trickling filters and the production of artificial wetlands are good solutions to small scale sewage treatment.
6. Both backyard and commercial composting reduce the need for large landfills for disposal of solid waste.

SUMMARY

I. Water Pollution by Microorganisms and Chemicals
   A. Contamination of water with pathogenic organisms remains a major cause of epidemic diseases.
   B. Chemicals contaminating water supplies caused more than a fourth of all reported outbreaks.

II. Water Testing and Treatment
   A. To ensure its quality, water is treated in various ways to remove contamination, especially by pathogenic organisms.

III. The Microbiology of Waste Treatment
   A. Microorganisms degrade the components of sewage to inorganic compounds.
   B. Microorganisms reduce the biological oxygen demand (BOD) of sewage.
   C. Methods for waste treatment
      1. Primary treatment is designed to remove materials that will settle or sediment out.
      2. Secondary treatment is designed to stabilize most of the organic materials and reduce the BOD of the sewage.
      3. Tertiary treatment is intended to remove nitrates and phosphates.
   D. Digester treatment involves the action of anaerobic organisms on solids remaining in sewage after aerobic treatment.
   E. Pathogenic bacteria are generally eliminated from sewage during secondary treatment except for some disease-causing viruses.
   F. Sewage can also be treated on a small scale using lagooning, trickling filters, septic systems, and artificial wetlands.
   G. The use of treated waste residue is still a problem.

IV. Microbiology of Solid Waste Treatment
   A. Sanitary landfills have been a low-cost method for disposing of solid waste.
   B. Backyard and commercial composting are an attractive alternative to landfill.

VOCABULARY: TERMS AND DEFINITIONS

The following list contains new terms introduced in this chapter. Use these terms to fill-in the blanks of the sentences that follow and you will have a definition or description of each new term.

trickling filters	**septic**	**biological oxygen demand**
activated sludge	**lagooning**	**stabilization**
artificial wetlands	**secondary**	**digester**
mineralization	**sanitary landfill**	

1. The mechanism by which populations of aerobic organisms stabilize sewage at most secondary treatment plants

    is called the _____ method.

2. _____ is the conversion by microorganisms of organic materials to inorganic

    forms. Another term for this process is _____ .

3. A method of sewage treatment that can be used by small communities and involves a series of ponds designed for

    aerobic and anaerobic stabilization is known as _____ .

4. _____ spray controlled amounts of sewage over beds of coarse gravel and rocks

    where communities of organisms degrade the sewage.

5. _____ treatment involves the chemical and biological processes of converting remaining organic
    materials in sewage into odorless inorganic substances that can be reused.

6. Within a sewage _____ , anaerobic organisms act on the solids through anaerobic stabilization.

7. The term _____ designates the oxygen-consuming property
    of a wastewater sample and is roughly proportional to the amount of degradable organic material present.

8. A _____ tank relies on anaerobic digestion and the proper distribution of the overflow
    into a drainage field.

9. In _____ sewage is channeled into shallow ponds where it remains for several
    days to a month.

10. The method of solid waste disposal where the waste is covered over each day with dirt

    is a _____ .

SELF-TEST OF READING MATERIAL

1. Which of the following was primarily responsible for the development of sewer systems?
    a. no more sites for landfills
    b. cholera
    c. plague
    d. community betterment groups
    e. odors

2. Everyday each American produces an average of _____ gallons of sewage.
   a. 1.2
   b. 12
   c. 120
   d. 1200
   e. 5

3. Which of the following treatments removes phosphates and nitrates from sewage?
   a. primary
   b. artificial wetlands
   c. secondary
   d. digester
   e. tertiary

4. What type of gas produced in a home digester can be used to heat homes and cook food?
   a. oxygen
   b. methane
   c. propane
   d. hydrogen
   e. carbon dioxide

5. During which of the following treatments are pathogens generally eliminated?
   a. primary
   b. artificial wetlands
   c. secondary
   d. digester
   e. tertiary

6. The biological oxygen demand roughly measures the amount of _____ in water.
   a. degradable organic material
   b. oxygen
   c. degradable inorganic material
   d. All of the above.
   e. Only a and c are correct.

REVIEW QUESTIONS

1. Describe the levels of sewage treatment. Do you know what level your community uses?

2. What is the BOD and how is it measured?

3. Contrast the activated sludge method and a digester.

4. How effective is each of the stages of sewage treatment in removing bacterial pathogens? Viral pathogens?

5. What are the disadvantages of a sanitary landfill?

6. How can composting help?

7. List the steps for making a compost pile at home.

ANSWERS:
   Vocabulary: Terms and Definitions
     1. activated sludge 2. stabilization/mineralization 3. artificial wetlands 4. trickling filters 5. secondary
     6. digester 7. biological oxygen demand 8. septic 9. lagooning 10. sanitary landfill
   Self-Test of Reading Material
     1. b 2. c 3. e 4. b 5. c 6. a

NOTES

NOTES

34 MICROBIOLOGY OF FOOD AND BEVERAGES

There is both good news and bad news concerning the activities of microorganisms in food and beverages. The bad news is that microorganisms are responsible for the spoilage of food. When this spoilage prevents food from being used in a world of many starving people, the loss is tragic. We usually, however, measure the loss in terms of dollars and it can be extensive. The good news is that microorganisms are also responsible for many of our fine and delicious foods. In some respects this food production is actually controlled spoilage. Many of our fermented foods were produced initially because the fermentation also preserves them. Knowing what spoils food can teach us how to preserve it.

KEY CONCEPTS

1. A variety of microorganisms can use human food as a growth medium. The endproducts they produce can be desirable (fermented foods), undesirable (spoilage), or harmful (food poisoning).
2. Factors such as available moisture, pH, storage temperature, and intrinsic properties of the food product can influence the type of microorganisms that grow and predominate in a food product.
3. A variety of foods such as bread, cheese, and alcoholic beverages result from metabolic activities of microorganisms.
4. Food spoilage can be eliminated or retarded by altering the conditions under which microorganisms grow.

SUMMARY

I. Factors Influencing the Growth of Organisms in Foods
   A. The availability of water plays a role in determining which microorganisms can grow.
   B. Low pH inhibits many bacteria.
   C. Low storage temperature inhibits the growth of spoilage microbes.
   D. Intrinsic factors such as available nutrients, protective coverings, and naturally occurring antimicrobial substances influence microbial growth.

II. Microorganisms are used in the production of food and beverages.
   A. Some milk products are made using lactic acid bacteria.
      1. Yogurt
      2. Cheese
   B. Pickles, sauerkraut, and olives are also made using lactic acid bacteria.
   C. Soy sauce is traditionally made by fermenting soy beans.
   D. Bread is produced with the yeast *Saccharomyces cerevesiae*.
   E. Alcoholic beverages are produced by fermentation yeasts such as *Saccharomyces*.
      1. Wine-grapes and other fruits are fermented to produce wines.
      2. Beer is made by fermenting starches.
      3. Distilled spirits
      4. Vinegar is made by *Acetobacter* species oxidizing alcohol to acetic acid.

III. Food Spoilage
   A. Bacteria commonly spoil fresh foods.
   B. Fungi commonly spoil breads, fruits, and dried foods.

IV. Food-borne Illness
   A. Food-borne intoxication occurs when a microorganism growing in a food produces a toxin.
   B. Food-borne infection occurs when living organisms in a food product are ingested and colonize the intestine.

V. Food Preservation

VOCABULARY: TERMS AND DEFINITIONS

The following list contains new terms introduced in this chapter. Use these terms to fill-in the blanks of the sentences that follow and you will have a definition or description of each new term.

lyophilization	**wort**	**koji**
amylase	**rennin**	**malting**
curd	**whey**	**brine**
water activity	**hops**	

1. During the production of soy sauce, the initial mixture of salt soybeans, wheat and the mold, *Aspergillus oryzae*,

   is called ————————————————.

2. The ————————————————— is a measure of the amount of moisture available in foods.

3. The enzyme added to milk during the production of cheese to cause it to coagulate is called ——————————.

4. To preserve vegetables, they are pickled in ——————————————.

5. The dried flowers of a vine-like plant that are added to beer and produce the typical bitter taste are

   called ————————————.

6. ———————————————— is another term for freeze drying.

7. During beer making, the conversion of the barley starch to sugars is called ——————————————.

8. In the making of cheeses, following the treatment of the milk with an enzyme, the coagulated protein, called

   the ——————————— , is separated from the liquid portion, called the ———————————————.

9. After beer has been malted, the germinated grain is added to warm water to give an extract called ——————————.

10. The starch digesting enzyme is called ——————————————————.

SELF-TEST OF READING MATERIAL

1. Which of the following is essential for the growth of all microorganisms in food and beverages?
   - a. vitamins
   - b. oxygen
   - c. water
   - d. sugars
   - e. darkness

2. Bacteria with an optimum growth temperature around $10^{\circ}C$ are capable of causing spoilage in a refrigerator and are called
   - a. thermophiles
   - b. mesophiles
   - c. acidophiles
   - d. psychopathic
   - e. psychrophiles

3. Which group of bacteria is most likely to spoil a freshwater trout caught in a cold mountain stream and preserved in salt?
    a. psychrophiles
    b. halophiles
    c. anaerobes
    d. thermophiles
    e. rockfordphiles

4. Salts and sugars preserve foods because they
    a. make them acid.
    b. produce a hypotonic environment.
    c. deplete nutrients.
    d. produce a hypertonic environment.
    e. provide nutrients.

5. Acidic foods are more likely to be spoiled by
    a. bacteria.
    b. fungi.
    c. acid sensitive organisms.
    d. thermophiles.
    e. Acid foods are rarely spoiled.

6. Sulfur dioxide is added to wine during its production to enhance the growth of *Lactobacillus*.
    a. True
    b. False

7. The thermal death time for determining sterilizing conditions, takes into consideration
    a. the temperature.
    b. the time.
    c. standard conditions of killing.
    d. the organism.
    e. All of the above are taken into consideration.

8. For proper pasteurization, ice cream would have to be pasteurized longer and at higher temperatures than milk.
    a. True
    b. False

9. The colors and flavors of red wine come from
    a. the grape skins and seeds.
    b. distillation.
    c. natural additives.
    d. the yeasts used.
    e. natural and artificial additives.

10. Vinegar production requires
    a. aerobic conditions.
    b. ethyl alcohol.
    c. anaerobic conditions.
    d. a and b are correct.
    e. b and c are correct.

REVIEW QUESTIONS

1. List the factors that influence the growth of microorganisms in foods and beverages. How are these factors any different from those that influence microbial growth in general?

2. Describe the three methods of pasteurization.

3. List the four categories of food-borne illnesses and give examples.

4. If your jar of pickles spoiled, would you expect to isolate fungi or bacteria? Why?

5. What factors impact on the proper times and temperatures that foods must be subjected to in order to kill the microorganisms present?

6. When foods are salted, smoked or sugared, what actually prevents the growth of unwanted microorganisms in each case?

ANSWERS:
Vocabulary: Terms and Definitions
1. koji 2. water activity 3. rennin 4. brine 5. hops 6. lyophilization 7. malting 8. curd/whey 9. wort 10. amylase
Self-Test of Reading Material
1. c 2. e 3. b 4. d 5. b 6. b 7. e 8. a 9. a 10. d

NOTES